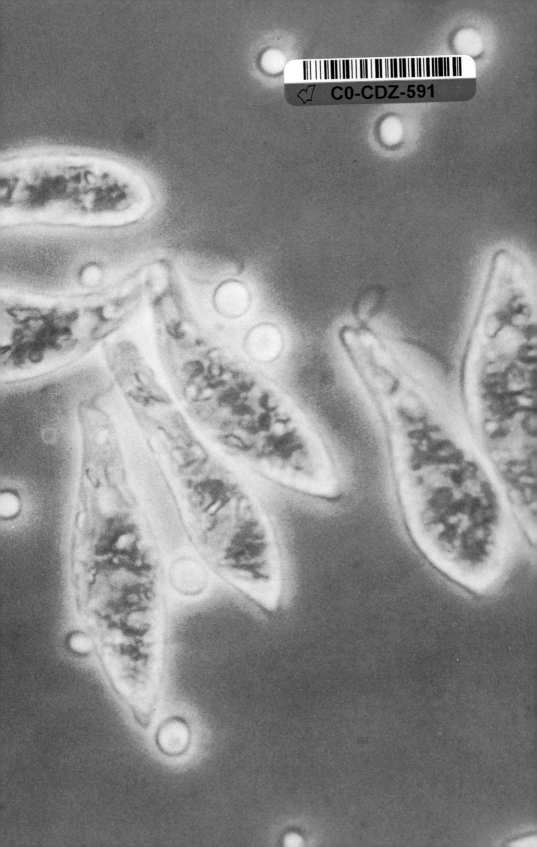

THE MARVELOUS ANIMALS

The Natural History Press, publisher for The American Museum of Natural History, is a division of Doubleday and Company, Inc. Directed by a joint editorial board made up of members of the staff of both the Museum and Doubleday, the Natural History Press publishes books and periodicals in all branches of the life and earth sciences, including anthropology and astronomy. The Natural History Press has its editorial offices at The American Museum of Natural History, Central Park West at 79th Street, New York, New York 10024 and its business offices at 501 Franklin Avenue, Garden City, New York.

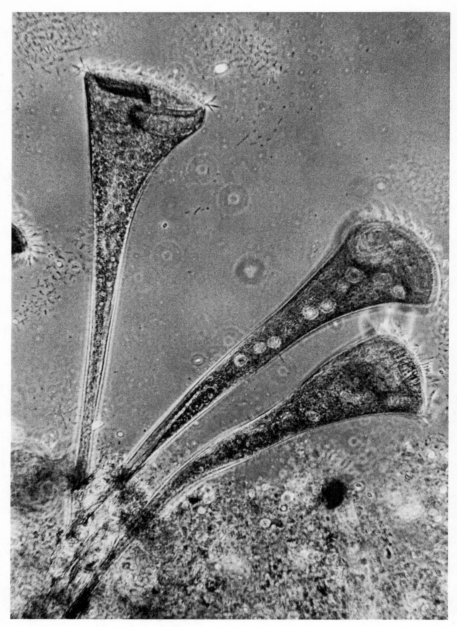

FRONTIS: A group of stentors extended for feeding.
Photograph by Eric Gravé.

THE MARVELOUS ANIMALS

AN INTRODUCTION TO THE PROTOZOA

HELENA CURTIS

Drawings by Shirley Baty

PUBLISHED FOR

The American Museum of Natural History

THE NATURAL HISTORY PRESS

GARDEN CITY, NEW YORK

ISBN: 0-385-05731-8
Library of Congress Catalog Card Number 67–15370
Copyright © 1968 by Helena Curtis
All Rights Reserved
Printed in the United States of America
9 8 7 6 5 4

To My Mother and Father

This book has been a pleasure to write and several people have helped to make it so. Joseph A. Mazzeo of Columbia University introduced me to the protozoa in the first place, and his skills and enthusiasms as an amateur protozoologist have been a constant source of assistance and encouragement. William Trager of the Rockefeller University permitted me to attend his lectures, which were invaluable, and knew the answers to all my questions. John Lee of The American Museum of Natural History read the manuscript and suggested a number of useful changes and additions.

The parts of this book that deal with Gonyaulax and Tokophrya appeared first in a somewhat different form in *The Rockefeller University Review*. They are included here with the kind permission of the publisher. I am particularly grateful to J. W. Hastings, now of Harvard University, and to Maria Rudzinska of the Rockefeller University who have done much of the work reported here on these organisms and who helped me with the stories as they first appeared in the *Review*.

I would also like to thank Mark and Sally Curtis, not because they make writing any easier—which they don't—but for their unfailing liveliness and good cheer.

H.C.
April 1967

CONTENTS

Contents

Contents

LIST OF ILLUSTRATIONS

List of Illustrations

THE MARVELOUS ANIMALS

CHAPTER I

"ALL ALIVE IN A DROP OF WATER"

"This was for me, among all the marvels that I have discovered in nature, the most marvelous of all," wrote Antony van Leeuwenhoek, "and I must say, for my part, that no more pleasant sight has come before my eye than these many thousands of living creatures, seen all alive in a drop of water, moving among one another, each several creature having its own proper motion." Leeuwenhoek, a lens-maker from Delft, first discovered the protozoa in 1675, and from that day onward peered at them through his tiny, improbable microscopes with undiminished delight until his ninetieth year.

And a pleasant sight they are indeed. Their shapes range from teardrops to bells, barrels, cups, cornucopias, stars, snowflakes, and radiating suns, to the common amoebas, which have no real shape at all. Some live in baskets that look as if they were fashioned of exquisitely carved ivory filigree. Others use colored bits of silica to make themselves bright mosaic domes. Some even form graceful transparent containers shaped like vases or wine glasses of fine crystal in which they make their homes.

There are more than thirty thousand species of protozoa living today, everywhere around us, billions and billions of individuals, far more than all the other animals in the world combined. In one small pond or even in a single culture flask, the various creatures will inhabit their own microcosms, the green, chlorophyll-bearing plantlike forms on the surface scum, amoebas crawling on the bottoms, some species clinging to the sides, others swarming around bacteria or fastened to bits of plants or other animals. One type of protozoa lives on mountain peaks, forming a red film on the

1

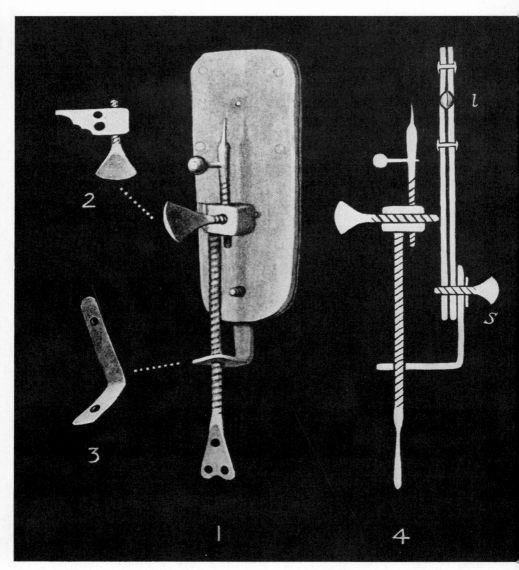

FIGURE 1 Leeuwenhoek's microscope. The minute biconcave lens, marked l
in the diagrammatic side view (4) was mounted between two thin plates of
brass and the object to be examined was held in front of the lens on the
point of the rod. In the case of the protozoa, the specimens were first
mounted between two thin glass plates or placed in fine capillary tubes. The
position of the specimen and the focus of the lens could be adjusted by the
various screws. In order to see anything, it would have been necessary to
place the eye almost in contact with the lens. Photograph reproduced with
permission from *The Unseen World,* René Dubos, 1962, published by The
Rockefeller University Press.

snow, another has been found in the near-boiling waters of hot springs, another in the ice floes of the Arctic. Even the desert has its own protozoa, which spend most of their existence in dried shells, coming to life perhaps only once a year for an hour or two following a rainfall.

Protozoa are found in the fresh waters of mountain streams, in brackish swamps, in rain barrels, and in sewers. Many live in the soil, about five or six inches below the ground, their entire universe the thin film of water between a few particles of earth. The most common and numerous of all inhabit the ocean waters forming much of the plankton, the rich meadow of the sea, on which the whole chain of marine life depends. Ocean travelers can recall nights when the ship's wake shone for miles with incandescent light, and seaside-dwellers remember evenings when a wading child traced an enchanted moonpath through the water, the result of the flashing lights of millions of protozoa. Many common species are found the world over. A pond in California may yield the same protozoan population as a similar pond in England or China. The animalcules found by Leeuwenhoek in the small inland lake near Delft three hundred years ago are likely to be there today and in lakes on every continent. Frogs all over the world carry similar protozoan parasites in their intestines; the sea urchins from every shore of every continent are hosts to like species. Malaria parasites have been found from ancient Greece to the twentieth-century Viet Nam marshes, wherever they can set up their curiously complex life cycle requiring both men and mosquitoes.

One reason for the ubiquity and hardiness of many of the protozoa is their ability to form cysts. Depending on the species, protozoa make cysts from many different materials, including cellulose, silica, and sometimes a chitin-like substance related to the tough protein from which insects make their exoskeletons and lobsters their shells. Some protozoa encyst to reproduce while others encyst after a large meal, withdrawing in a sort of postprandial lethargy. Many retire into cysts when food becomes scarce or in the face of cold or heat and drought. In this state, they may be carried by the winds for miles, across oceans and continents, or they may remain dormant in the same spot, clinging to a little germ of life and waiting to be awakened.

3

WHAT ARE THE PROTOZOA?

How do we define the protozoa? One protozoologist states firmly in the introduction to his text that a protozoan is anything that he includes in his book. Another expert, asked the same question, thought a minute and said, "If it wiggles. . . ." Yet, despite this apparent lack of dogmatic guideline, protozoologists are in virtually complete agreement about the thousands of species classified as protozoa, although not always in accord with the details of their classification. The difficulty in defining the protozoa, which means "the first animals," arises because scientists concerned with "the first plants" also study some of the same species, classifying them with the algae. This overlapping occurs not because of any real difference of opinion but because on the one-celled level of existence there is no sharp dividing line between plants and animals. Biologists are now tending to group both protozoa and algae together as protista—a large group that is considered a separate "kingdom" (the many-celled plants constituting a second kingdom and the many-celled animals making up the third). The protista include not only protozoa and algae but other quite different unicellular organisms, such as bacteria and fungi.

So, in short, protozoa are the creatures we talk about in this book. All of them are one-celled. Some of them are clearly animals; that is, they do not possess chlorophyll or any similar pigment and are therefore not capable of photosynthesis. Others are included that do possess chlorophyll and do photosynthesize. These wiggle, at least most of them do, and all of them are so closely related to animal forms that separating the two groups becomes impossible. Examples of this impossibility can be found in the chapter on the flagellates, Chapter VII.

The reason for this taxonomic dilemma can be more clearly understood if you recall that ideally the classification of living things reflects their evolutionary history. And most of the evolutionary history of the protozoa still remains obscure. There is probably no one single common ancestral form, and many of the links between the major groups will probably never be reconstructed. For instance, it appears as if certain algae lost their chlorophyll and joined the protozoan, or animal, world quite late in their evolutionary history, and it is not impossible that certain protozoa may have acquired photosynthetic capacities, becoming algae, at least to the botanists. Some suggestions as to how such phenomena can occur will be found in the pages that follow.

A small avant-garde of protozoologists are now studying the biochemistry of the protozoa, the composition of various cellular components, the nutritional requirements of different species, and the way in which different types assimilate food, burn, store, and use it for growth and reproduction. Such studies are beginning to reveal affinities between various groups and thus provide clues to the problem of protozoan genealogy and systematics.

UNDER THE MICROSCOPE

For those whose original introduction to the one-celled animals was a sketch in a biology textbook, the first sight of the living animal is a revelation. Drawings are useful, of course; they pick out the significant features, often almost indistinguishable to the amateur eye, and they are invaluable for classification. Some of the drawings, particularly those made by the early microscopists, are also works of art. But the usual textbook drawing bears little resemblance to the real animals or rather about as much resemblance as does, for example, the bare floor plan of your home compared to how it seems with all the family in it, a fire in the fireplace, and supper cooking on the stove. None conveys the fragile transparency of the protozoa, the shimmering surface through which one can watch the flow of cytoplasm, the movement of organelles, and the motion of the incredibly swift appendages. Most protozoa are almost as transparent as the water in which they live. Those containing chlorophyll are a sparkling green, and some are shining and iridescent shades of blue, rose, or yellow.

They range almost as widely in size as in shape. Single cells may be as small as two microns (a micron is 1/25,000th of an inch and is the usual unit of measurement for the protozoa) to several millimeters in diameter. The largest are visible to the unaided eye, and all but the smallest can be observed with simple instruments.

Like all cells, protozoa contain a nucleus and cytoplasm. Some of them are surrounded by a firm cell wall such as is found in most plant cells. Others have shells or skeletons. In some, the outermost layer of cytoplasm is hardened into a protective coat, the pellicle. Others, like the common amoeba, are separated from the world in which they live by a membrane so fine as to be invisible. Within these varied outer boundaries can be found a number of intracellular structures that will be described in the chapters that follow.

THE SCIENCE OF PROTOZOOLOGY

Two hundred years ago, almost every gentleman had his own microscope, as much a mark of his cultivation and interest in the world around him as the richly bound volumes that lined the walls of his library or the handsomely mounted globe that stood by the window leading to the garden. In one melodrama of the eighteenth century, the young heroine regretfully declines the pleadings of an importunate suitor who wishes to carry her off with a cry of, "What! and leave my Microscope!" So protozoology, although not always demanding of such a noble sacrifice, began and has remained first and foremost a pleasure and a delight.

There are, of course, other reasons for studying these marvelous animals, more practical and more "scientific." At about the time that Pasteur and Koch established that bacteria cause disease, the protozoa came to be associated with amoebic dysentery and with malaria, and since that time the parasitic protozoa have been investigated for medical reasons. Some of the diseases for which these parasites are responsible constitute serious medical problems, but actually, considering their enormous number and ubiquity, the one-celled animals are surprisingly harmless. In fact, they may be considered much more as friends and allies. They are a basic part of the long food chain to which we contribute very little and on which our survival depends, as well as that of the rest of the animal world. The protozoa are also our major assets in the weakly waged struggle against the pollution of our natural waters (a single paramecium can devour five thousand bacteria a day). But the study of the effects of man-made contaminants—pesticides, for example —on the fate of these useful protozoa has hardly begun, and the first reports are disquieting. Also, the one-celled animals possess the ability to select certain chemicals from the environment in which they live, an ability essential to their nutrition. By this same mechanism, however, some types of marine protozoa have been at work over the last two decades storing away strontium 90 and other radioactive by-products, which are accumulated by the fish and animals that eat them and are eaten in turn, in a long chain leading to man, the ultimate consumer.

THE PROTOZOA AS CELLS

"No one will doubt," wrote Leeuwenhoek, "that these . . . animalcules are provided with organs similar to those of the higher animals. How marvelous must be the visceral apparatus shut in such animalcula." These words foreshadowed a controversy which only ended in our own time. So complex are these "little animals" that a century passed before it could be decided whether or not they were "simple cells."

The Dane, O. F. Müller, was the first great taxonomist of the protozoa, describing and naming more than one hundred species during the closing decades of the eighteenth century. He grouped them among the *Infusoria,* which he regarded as a class of worms, but which comprised many heterogeneous animals, including a number of molluscan larvae, microscopic plants, and small many-celled animals. Müller regarded all these diverse creatures as the simplest of all living beings, composed of some sort of homogeneous material.

Some fifty years later, the German microscopist C. G. Ehrenberg, using the finest microscopes of his time—indeed, comparable to the microscopes of our own day—reported seeing within the protozoa complete systems of minute but perfect organs: eyes, muscles, stomachs, ovaries, testes, and lungs. His report on the subject, "Infusorial Animalcules as Complete Organisms," appeared in 1838. By coincidence, this was the same year in which M. M. Schleiden

7

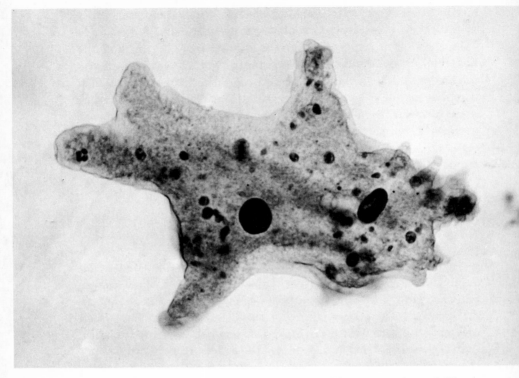

FIGURE 2 Amoeba, a "primitive living jelly." (Up to 600 microns.) Photograph courtesy Ward's Natural Science Establishment, Inc. (*Note:* in this and succeeding captions, figures given in parentheses are approximate sizes of animals.)

first proposed that all plants were made up of independent, separate units, or cells, each containing a nucleus. In the following year, Theodor Schwann extended Schleiden's cell theory, hypothesizing that all animals, as well as all plants, were made up of aggregates of single cells.

The French biologist, Félix Dujardin, was the first to recognize the protozoa as a special group. His systematic treatise, published in 1841, laid the basis for the modern classification of the phylum. Dujardin, who studied in particular the amoeba-like organisms (a fact that certainly influenced his ideas on the nature of the living material), was the coiner of the word *sarcode,* which he used to describe the material that could be squashed out of the living cell— *"une substance glutineuse, simple, et homogène,"* which he viewed as a sort of "primitive living jelly," common to embryos and simple forms of life.

By 1850, the idea of evolution was in the air and the word *protoplasm,* meaning "first formed," came into use, carrying with it the connotation of a substance of universal identity in *all* cells, a crucial concept for biology. By 1869, the great biologist Ernst Haeckel, who also began his career as a protozoologist, was able to make the important generalization that the protozoa are one-celled animals and the metazoa—or many-celled animals—are composed of groups of cells that derive from a single cell, the fertilized egg, by successive cell divisions.

Yet, until very recently, some microscopists proposed that the protozoa be called not unicellular but "acellular." While they were able to recognize, as Ehrenberg could not, that the protozoa were not minute metazoa, they still felt that anything as complicated as these extraordinary animalcules could not possibly be "simple cells."

Today modern techniques of microscopy—particularly the electron microscope—and of biochemistry have solved the problem: there is no such thing as a simple cell. All cells are enormously complex, an intricate arrangement of structures and forms, some large enough to be seen with the light microscope, others almost as small as a molecule, but each having a special function and a configuration that is intimately linked with the carrying out of this function. The *sarcode* exists no more; even the word "protoplasm" has fallen into disrepute and has been written out of many modern textbooks. It remains a handy term for the varied contents of the cell, however—there is no other word—and also a useful reminder that all life, no matter how varied in form, is composed in essence of the same richly textured fabric.

In short, the cells—whether the cells of our own body or of the single-celled animals—are no longer envisioned as mere building blocks or surviving droplets of the primordial ooze. They are, architecturally, enormously complex. With this new knowledge, the protozoa have achieved a new "usefulness" to fundamental biology.

9

Each protozoan cell contains the features common to all cells: mitochondria, nuclei, chromosomes, membranes, lysosomes—all constructed along the same general design, repeated over and over throughout all forms of life. These structures, which the protozoan cell shares with the cells of higher animals, can often be more satisfactorily studied biochemically and microscopically in the protozoa, for they can be observed in their natural environment and not in the artificial conditions of tissue culture.

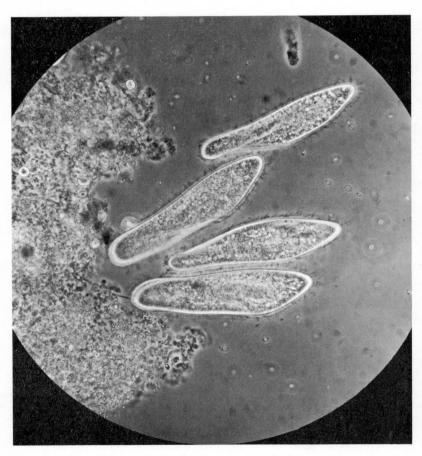

FIGURE 3   Paramecia were the slipper animalcules of the early microscopist. (150 microns.) Photograph by John J. Lee.

There are differences, of course, between the protozoan and the metazoan cell. The protozoa are complete organisms and must do by themselves all that is essential to maintain their own life and that of their species. Any given type of human body cell has many of its functions performed for it by other types of cells and therefore does not require some of the structures which the free-living protozoa possess. Yet some of these unusual structures have persisted in certain types of body cells, like the flagellum in the human sperm cell. The protozoa provide a rich bonus for those who choose to explore and speculate upon the outer limits of the potentialities of a single cell.

To the biologist, another great advantage of the protozoan cell, besides its fascinating complexity of structure, is the rapidity and frequency with which these structures are generated or transformed. Before every cell division, after every conjugation, upon each emergence from a protective cyst, the protozoan cell may undergo a spectacular morphogenesis. The microscopist, for example, can watch the creation of an entirely new set of organelles during division, organelles more complex than any found in the specialized cells of metazoan tissues. Now that biology is ready to begin the study, on the microscopic level, of how living creatures take their shapes, the protozoa are coming to be recognized as ideal experimental animals for such investigations.

EXPERIMENTERS IN EVOLUTION

Finally, the protozoa excite our curiosity because they afford a glimpse of the world of billions of years ago. Their ancestors were early experimenters in living, and with them the most fundamental problems had to be worked out: how to get food; how to assimilate it; how to store it; how to eat and digest another animal, even its twin, without somehow accidentally digesting itself; how to repair and co-ordinate all of the necessary machinery; how to reproduce its own kind, providing the right balance between genetic continuity and diversity.

A visible index of the extent of this experimentation is the variety of the protozoa now in existence and the extraordinary range of environments they can live in. The protozoa tried everything and in so doing adapted themselves to survival in the most improbable of situations, ecological niches whose possession is virtually uncontested by other living things. Only the bacteria live in more varied and improbable habitats. The great prolificacy of the one-celled animals and the long span of time during which they have evolved —as they are evolving even today—provided an almost unequaled opportunity both for great variation and for the forces of natural selection to pick and choose, editing out mistakes and reinforcing the successes. Many of the records of this long, ruthless experiment have been lost forever. Nor are the living protozoa in any way ancient. They are as modern as you and I; in fact, any protozoan you are likely to see was born yesterday if not more recently. Yet, there are clear indications some forms have remained virtually unchanged through millenniums, since long before the thunder lizards ruled the earth or man made his first and still tentative appearance on this planet. Surviving in the protozoa of today are suggestions of how it might have been.

CHAPTER II

"EACH ITS OWN PROPER MOTION"

When one first looks through a microscope into the protozoan world, one is struck by the seemingly ceaseless activity of its inhabitants. They exhibit a fantastic variety of motions—darting, swimming, spinning, scurrying, bouncing (no less) through their microcosm. Even many of the so-called seated, or sessile, forms that anchor themselves to the bottom or to bits of vegetation or debris are constantly stretching and retreating, popping forward and back as if held by tight elastic bands, whirling the water around them and creating a constant inrush of food-bearing currents. These activities require a tremendous amount of energy— like an automobile driven full speed, day and night, not stopping a moment even for repairs. And because of these enormous fuel requirements, the protozoa in turn cannot afford to relax their activities, which are, in essence, a constant search for food.

Each major group of protozoa has distinctly different methods of locomotion. Or, to put it another way, the protozoa have been divided into separate major groups on the basis of their methods of locomotion. This is very fortunate for the amateur, as it enables him, almost from the beginning, merely to glance at one of these busy little specks of protoplasm and immediately assign it to its appropriate category.

AMOEBOID MOTION

Perhaps the most familiar of all types of protozoan motion is that of the amoeba, which moves itself along by constantly changing its shape. Some two hundred years ago, the commonest member of this family was given the name of "little Proteus" after the sea god of Greek mythology who eluded capture by changing his form. Later in the course of the endless revising and name-changing that has marked so much research in protozoology, the name of "little Proteus" was changed to *Amoeba proteus* ("amoeba" from the Greek word meaning change).

The amoeba captures the imagination because it seems to represent the simplest form that life might take—a shapeless bit of ooze into which some divinity (perhaps Proteus himself?) once breathed a vital spark. The amoeba changes shape and so moves forward because its protoplasm is able to assume two different consistencies. One—the gel—is quite rigid, like a well-set gelatin; and the other —the sol—is more liquid. In an advancing foot (the pseudopodium) of the amoeba, the innermost contents, in the form of sol, flow forward through the semirigid tube (or cylinder) formed by the outermost contents (the gel). This you can see readily even under a low-powered microscope, since *Amoeba proteus,* which is very large as protozoa go—up to six hundred microns or more when elongated—is filled with many small granules that travel along with the moving sol and so mark its currents. At the advancing end of the pseudopodium, the sol streams forward and then falls back fountain-like onto the surrounding gel. As it joins the gel, it hardens and so forms an extension of the tube. All of this has been observed countless times, yet no one is quite sure how it actually makes an amoeba go forward. Some say (and this is the easiest way to think about it) that it is just like squeezing out a tube of toothpaste—from the bottom, that is. Others feel that the contraction of the protoplasm at the advancing end of the pseudopodium pulls the rest of the amoeba after it. Still another group holds that the motion depends upon the interaction between the contracted molecules making up the gel and the more expanded ones of the sol, so that the one crawls along the other, like a system of interlocking gears. There is evidence that this is the way in which some of our own muscular movements take place, but the question has yet to be answered for amoebas.

Watching these large amoebas, one sometimes feels that they do not know where they are going. One will stretch out two great

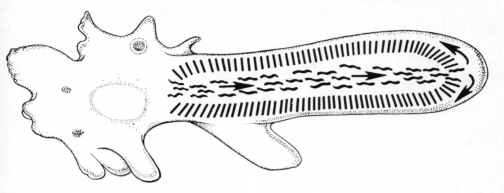

FIGURE 4 In the amoeba, the liquid sol in the advancing foot reaches the tip and flows back fountain-like on the surrounding gel.

pseudopodia simultaneously in almost opposite directions, seeming almost on the verge of pulling itself asunder, then seem to hesitate a moment, apparently resolve the conflict, and retract one of the pseudopodia with the sol streaming backward into its body followed by the dissolving gel, and then all rushing forward into the chosen pseudopodium. This is probably the amoeba's way of exploring its environment, for it has been shown clearly—in fact, one can see for oneself—that an amoeba can sense the presence of prey at some distance and move purposefully toward it or can respond to an irritating stimulus, such as certain kinds of light, and reverse its direction.

When viewed from the top, the usual vantage point of the microscopist, *Amoeba proteus* appears to be flowing forward all in a single plane, like a spreading inkblot erasing itself behind as it moves. But actually, side views show that its pseudopodia do not just spread along the surface but, clumsy and elephantine, may reach over and forward; part of proteus' progress is accomplished by these awkward half-somersaults. Some amoebas have shells, or tests, shaped like inverted bowls with openings at the bottom. These amoebas extend their feet—or whatever portions of themselves they are using for feet at that particular moment—through these small openings, raise themselves off the bottom and totter along with their shell wobbling on top.

FIGURE 5 Side view of amoeba shows that it does not simply spread itself along like an inkblot but instead may form large "false feet" (pseudopodia) by which it lifts itself off the ground.

Amoeboid motion is a matter of curiosity not only to protozoologists. Some of the most interesting cells in our own bodies—the white blood cells that serve as scavengers in our blood streams—propel themselves in the same way, and the protoplasmic streaming by which the various particles of many plant and animal cells are kept in constant circulation is based also on these same principles.

FLAGELLAR MOTION

Each of us owes our presence here on this green planet to the unfaltering beating of one microscopic flagellum. The flagellum of the spermatozoan is like a tail and propels the cell forward by lashing from behind. This must be an efficient method as judged by the number and increase of us vertebrates, but, curiously, only a few species of protozoa are known to use their flagella in this way.

The smallest of the flagellates usually have one or two, sometimes three, hairlike flagella—some so fine they are difficult to see without special staining. These are attached to the front of their small, smooth bodies, which are usually round or pear-shaped. Some appear to use the flagella simply as little arms, swimming through the water in a sort of breast stroke or dog paddle. But most flagellar motion is by no means this simple. One of the most interesting and carefully studied is that of euglena, a chlorophyll-containing flagellate. The most common species, *Euglena gracilis,* is quite small—about fifty microns long—but some species are several times larger.

In the microbiology texts, euglena is identified as spindle-shaped, but for those of us now some generations removed from this admirable handicraft of spinning, it is more informative to describe it as shaped like a cigar or, in keeping with its function, like the hull of a submarine. One day, in fact, we may describe a spindle by explaining that it was euglena-shaped.

Euglena has two flagella arising at the base of an opening in its prow known as the reservoir. One of these flagella is very short, rudimentary in fact, and sometimes looks as if it were merely a bifurcated end of the other. The motion-imparting flagellum in euglena extends forward, out through the reservoir, and is quite long, about three fourths of the length of the body of the animal. As euglena moves through the water, this flagellum is held off to the side trailing a little behind it, and slightly coiled. About twelve times a second—much too fast to see by ordinary techniques—it beats with a whiplike pulse running from base to tip along the coils. This strong, rapid beat imparts motion to the cell—two types of motion in fact: one a simple spinning motion, like a top, and the other a gyrating motion, which causes the front end of euglena to describe a series of circles as it moves forward. If one draws an imaginary line along the course euglena is traveling, the posterior

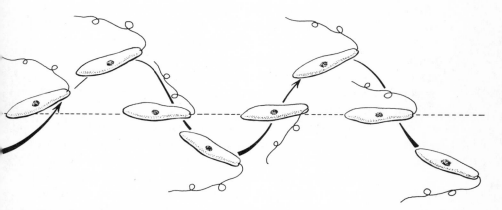

FIGURE 6 The trailing flagellum of euglena beats with a whiplike motion so that the body of the animal both swims and gyrates as it moves. The tail moves in an almost straight line; the head describes a series of wide circles. (40 microns.)

end of its hull moves almost unwaveringly along this line while the front end goes round and round it, never touching it. Those who understand the principles of hydrodynamics explain that this spinning-gyrating motion produced by the flagellum, which can really be thought of in this context as a sort of outboard motor, causes euglena to move like a propeller blade through the water.

The flagellum of euglena has all along its length a single row of fine hairs called mastigonemes. Microscopists before the end of the century had reported seeing mastigonemes, but until quite recently there had been a great deal of argument as to whether or not they really existed. Since it is necessary to stain and fix flagella before the mastigonemes can be seen under the light microscope, many microscopists contended that the fine hairs were actually a man-made result, the effects of fraying the delicate membrane of the organelle. Now, however, by the use of the electron microscope, four basic types of flagella have been distinguished: the simple flagellum, which is without mastigonemes; the acroneme flagellum, with a single terminal thread of variable length; the pantoneme flagellum, with two rows of mastigonemes along opposite sides; and the stichoneme flagellum, with a single row, such as that seen in euglena. Different types are seen very consistently among different species of flagellates—which is one of the reasons that microscopists are now sure that they actually exist—and they are now being used as a taxonomic guide.

Peranema is closely related to euglena and resembles it in appearance, with a hull-shaped body and a single flagellum protruding from the reservoir. It moves quite differently, however, from its slowly gyrating relative. The single flagellum is held stiffly out in front of it like a long pole, and just at the very tip of this pole a short, whiplike segment beats steadily and rhythmically. It is generally agreed to be impossible for this beating alone to move peranema so smoothly and efficiently, but it has also been shown that if this flagellum is removed or accidently lost all motion will stop until it fully regenerates. Peranema has a second flagellum which also arises at the anterior end but curves back along the underside, or ventral surface, of the body, to which it is firmly attached. Its function is not understood at all. Peranema is known to secrete a type of mucus, and some protozoologists believe that

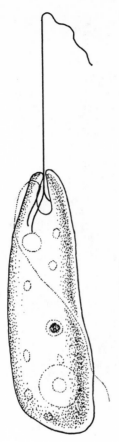

FIGURE 7 Peranema advances with a gliding motion, its anterior flagellum held stiffly in front of it. (40 microns.)

this may be the secret of its motion, envisioning the animal as actually skating or sliding along the slippery track it lays down for itself.

A quite different flagellar arrangement is seen in a group known as the dino-, or spinning, flagellates. The dinoflagellates are found in fresh and brackish waters the world over but occur in by far the greatest number in the oceans, where they make up a considerable portion of the plankton, the little floating wanderers of the sea. One typical dinoflagellate, gonyaulax, for example, has two flagella, one of which lies beltlike in a deep groove, or girdle, that runs all the way around the center of its body. Another groove, the

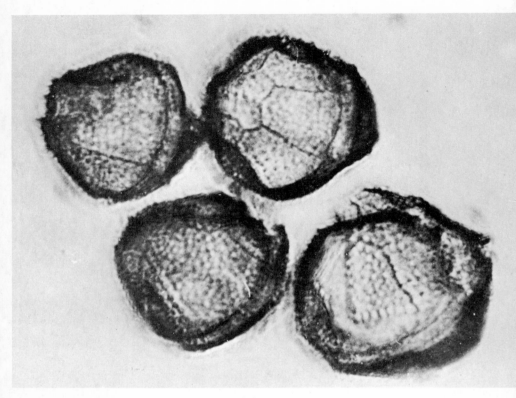

FIGURE 8 In *Gonyaulax polyedra,* a dinoflagellate, the flagella beat in the two grooves visible in the picture, causing the cell to spin like a top. (50 microns.) Photograph courtesy University of California, San Diego, Scripps Institution of Oceanography.

sulcus, runs at a right angle from the girdle along the underside of the cell to its base, and in this sulcus arises another, much smaller, flagellum. Each beats in its own groove, one imparting a forward motion, apparently, and the other a spinning action. These creatures are small and difficult to observe since they move in and out of the field of focus rapidly, whirling like tiny tops.

The novice, who sets out with a feeling of faint superiority toward his seniors who have observed these motions for so long without being able to explain them, quickly retreats to a position of humility and admiration that any details of these actions could have been recorded. And in this case, the great advances in microscopy in our generation are of little help, for once the creature has been stained and dried and fixed for viewing in the electron microscope, all motion has, of course, long ceased and one is left merely with the dead structure whose function can be only a matter of conjecture.

CILIATES

A third major class of the protozoa are those which move by the activity of their cilia. There is no precise definition of the difference between a cilium, which is the Latin word for eyelash, and a flagellum except that it is generally understood that a flagellum is much longer—up to fifty times as long as a cilium. Also, most free-living flagellates have only a few flagella, about four at most, while the ciliates are often covered with thousands of cilia arranged in orderly patterns.

The most familiar of the ciliates are the paramecia, the "slipper animalcules" of the early microscopists, and probably the most extensively studied among paramecia is *Paramecium caudatum,* which is large and common. *Paramecium caudatum* is covered with some twenty-five hundred cilia arranged in parallel rows. The beating of the cilia is synchronized, like the strokes of oarsmen, and the ciliary motion is in principle not unlike that of an oar, with a strong forward stroke and a limp recovery stroke. The oars of paramecium do not all stroke at the same time, however, but instead the stroking motion runs in waves down the animal's body, like wind rippling a field of wheat. In paramecium, these ripples of ciliary motion do not travel directly from front to rear along the axis of the animal, but follow a slightly oblique angle, slantwise down its body. For this reason paramecium normally rotates as it swims, if you give it enough room under the cover slip when you are looking at it through the microscope. All species of paramecia always rotate to the right whether they are swimming forward or backing up.

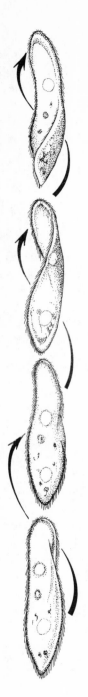

FIGURE 9   Paramecium rotates as it swims.

The cilia of paramecium are normally in a state of continuous movement. In a culture in which there is no food, paramecium may move as fast as two or three millimeters a second (this is equivalent to a car, reduced to protozoan scale, going about one hundred miles an hour) and maintain this speed for hours, but ordinarily it moves at slower speeds and, in bacteria-rich cultures, may spend much of its time browsing. It can reverse its ciliary motion and go backwards in an avoidance movement, or it can stop its ciliary beat when it encounters a solid object that requires exploration. This, is, of course, in contrast to the behavior of more familiar animals that remain still or at rest until they are stimulated or excited; technically speaking, paramecium is said to be under inhibitory rather than excitatory control.

Paramecia belong to a group of ciliates known as the holotrichs, which simply means that they have uniform cilia (trich is from the Greek word for hair) usually distributed all over their bodies. Another major group of the ciliates is known as the spirotrichs. In the spirotrichs, which represent in many respects the most complex and most highly evolved of the protozoa, groups of cilia are often fused together in clumps called cirri. These cirri, which are stiff structures, can be used like paddles for swimming but are more often used for walking. *Euplotes patella,* for example, which is common in standing fresh-water pools and ponds, has seventeen cirri on its ventral surface; most of these can be seen when looking at the animal from above—since euplotes is virtually transparent—although it is difficult to observe their activity from this vantage point. From time to time, however, one is fortunate to see euplotes from a side view—moving along a fragment of alga for instance—and then it can be seen that these cirri do indeed form quite sturdy legs on which the animal moves around in a very coordinated and purposeful fashion. As seen from above, euplotes and other ciliates that use their cirri for legs have typical scurrying motions, quite different from the rhythmic motions of the flagellates and the holotrichs—much more insect-like, or even terrier-like, creeping forward, snapping back, turning, constantly changing direction and speed as they explore the environment.

Not all the animals use their cirri in the same way. A particularly surprising one, halteria, for instance, a small spherical ciliate, has

FIGURE 10   A side view of euplotes. (90 microns.) Photograph by Eric Gravé.

some seven groups (depending on the species) of three or four cirri each, and by rapidly flicking its cirri halteria leaps in a series of highly erratic jumps. One gets to recognize this animal quickly, not because one ever really can look at it very long, but because of its highly characteristic bounce.

Cilia are found on many other cells besides those of protozoa; in fact, they have now been found in every known phylum, invertebrate and vertebrate, although in only very small organisms, such as the protozoa, can they be used for locomotion. Clams have cilia on their gills with which they beat in food particles; similarly in the frog, cilia help to propel small particles of food into the alimentary canal. In the fallopian tubes of the human female, the cilia of the epithelial cells lining their surface move the ovum toward the uterus. Our entire respiratory tract is lined with cilia, the beating of which serves to circulate a protective coat of mucus and also to ward off foreign particles breathed in with air. Cigarette smoke has been found to halt ciliary motion, and medical investigators have postulated that one of the first steps in the development of lung cancer may be this effect on the cilia, which permits cigarette smoke tar and other cancer-causing chemicals to light on the surface, the epithelium, of the respiratory tract.

Like the cilia of paramecium, these cilia of epithelial cells beat in waves. The direction of the beat is fixed. If a small square of cilia-bearing epithelium is removed from the pharynx and then reimplanted at right angles, the cilia will continue to beat but in a direction perpendicular to that of the others surrounding it.

Recently the electron microscope has revealed that all of these cilia, and all flagella as well, have the same internal structure, an orderly pattern known now in biological shorthand as the "nine-plus-two." Running through each flagellum or cilium from base to tip are a group of continuous longitudinal fibers. There are two central fibers, which are single, and nine groups of outer fibers, which always appear in pairs. Presumably, these fibers are contractile, the outer ones at least, and these contractions could well explain how cilia and flagella move. And each one of us—like all the other forms of life—share with the protozoa this same precise and delicate geometry.

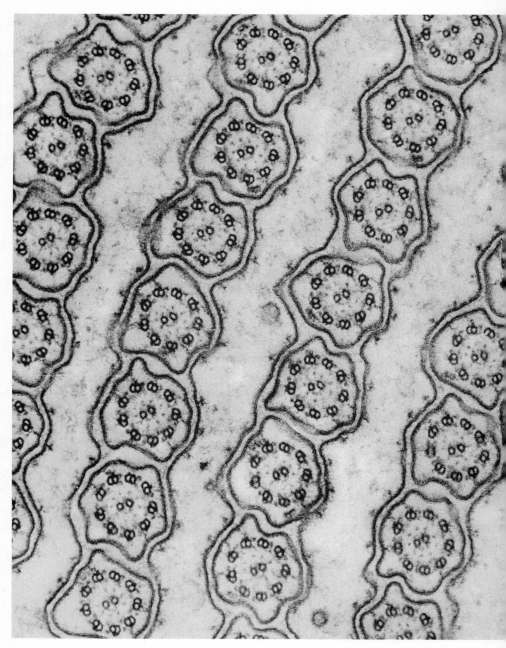

FIGURE 11   Electron micrograph showing the "nine-plus-two" structure characteristic of cilia and flagella. This is a cross section of the pellicle of trichonympha; a photograph of the living animal appears in Figure 53. Photograph
courtesy Dr. A. V. Grimstone.

CLASSIFYING THE PROTOZOA

One of the most famous instructors of protozoology once noted wearily that the protozoa are so much more attractive than their names, and indeed many an eager beginner has first opened a taxonomic manual only to close it again, almost immediately, when faced with the profusion of italicized names, often concocted from painful combinations of Greek and Latin. Even the expert cannot keep up. Several hundred new species are added each year, differing from those previously described perhaps by a single cirrus. This is not to say that the amateur should not eventually try, for example, to count the cirri of whatever one-celled animal he has found in his microdrop and to identify the creature by its particular or specific name. This is certainly the best possible training for the eye, and fun besides. But this is definitely not the way to begin. Fortunately, since the most general classification of the protozoa is based on obviously different methods of locomotion, the major groups are easy to recognize (except for some few aberrant forms). Start there. Those that move in amoeboid fashion are known as the sarcodines or rhizopods (meaning root feet); those that move by means of flagella are the flagellates, or magistophora, depending upon whether one prefers the Greek or Latin word for whip; and those that have cilia, during either all or part of their life cycle, are known simply as the ciliates. The fourth major group, all parasitic, are the sporozoa, which are distinguished to a large degree by having no special locomotor organelles at all for a major part of their life cycle. With practice the newcomer to microscopy will, as biologists say, "get his eye in" and soon begin to notice more and more of the special features of the creature he is observing. Then more precise classifications become possible and often useful. But the marvelous animals still remain far more attractive than their taxonomy.

CHAPTER III

WAYS OF LIFE

Early man managed to get a foothold in the world and to survive, a famous paleontologist once remarked, because he did not taste very good. As a consequence, although comparatively defenseless and a slow breeder, he could come down from the shelter of the trees, learn to stand erect, and develop an opposable thumb. By contrast, the protozoa are tasty indeed, and are consumed relentlessly by everything from the great whales to each other. And they in turn are enormous consumers, capable of devouring far more than their own weight in food in a single day—or even in a single meal—in their struggle for survival.

Protozoa feed in three different ways. 1) Substances dissolved in their watery environment may be taken in through the cell membrane. This method of nutrition, which is far more complicated than it sounds, is called saprozoic. 2) In the case of the plantlike protozoa, those that contain chlorophyll, photosynthesis is the chief energy source. 3) Others take in their food in solid particles, phagocytosis. Most of the protozoa make use of at least two of these methods, and some use all three.

SAPROZOIC NUTRITION

Saprozoic nutrition is both primitive and sophisticated. Most bacteria live saprozoically. The lineage of the bacteria is considered even more ancient than that of the protozoa. On the other hand, the cells of our own bodies derive their nourishment by this means, selecting their menu from the constant and varied supplies offered in the blood stream. The means by which a cell, any cell, can take in certain substances and exclude others is one of the most active areas of current research, since it is a crucial factor in cell physiology. All protozoa appear to receive at least some small amounts of vitamins and minerals in this way, and some, largely parasitic forms, depend entirely on this mode of nutrition. One of these latter is opalina, an especially handsome little animal that lives exclusively in the digestive tract of frogs. Opalina's ancestors had spent probably millions of years developing mouths—or cytostomes, as they are more properly called—before they moved into the frog, but the many species of this ubiquitous but harmless parasite have spent additional millenniums in losing all traces of mouth parts and now live an exclusively saprozoic existence.

The protozoa's need for materials in very small amounts is being exploited by biochemists for micromeasurements of nutrients. *Euglena gracilis,* for instance, requires vitamin B_{12}. In a medium in

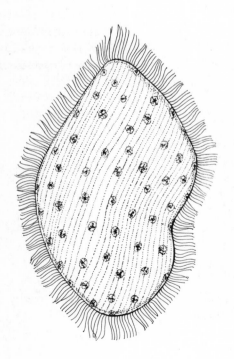

FIGURE 12   Opalina, which once had a "mouth," has lost it in the course of evolution and now absorbs its food from the digestive tract of the frog. (100–840 microns.)

which euglena has all the other vitamins and minerals it needs, its growth rate will be in direct proportion to the amount of B_{12} that is provided. B_{12} is present in the blood stream of normal, healthy persons, but lacking in the blood of patients with pernicious anemia. If a drop of blood from a patient suspected of having a B_{12} deficiency is properly prepared and added to the euglena culture in place of its regular B_{12}, the amount of growth of euglena will reveal quickly and accurately whether or not the patient has the disease and also the severity of it. This method is so sensitive that it can detect even a very small concentration (one part in a trillion) of the vitamin and so simple that it is being used routinely in many hospital laboratories.

PHOTOSYNTHESIS

The phytoflagellates, that group of protozoa that possess chlorophyll, usually depend on photosynthesis for their main energy source. Chlorophyll is a complex organic molecule that contains in its core a single atom of magnesium, much as the hemoglobin molecule is centered around an iron atom. Isolated in the test tube, chlorophyll is powerless, but if the molecules of chlorophyll are sandwiched between layers of protein and lipid they can convert carbon dioxide and water to glucose and oxygen. The production of each molecule of glucose requires 112,000 calories of energy from the sun and also a number of very complicated chemical steps, many of which have not yet been clearly charted. This process is obviously profligate in terms of calories expended, but the sun is a boundless source, and photosynthesis enables living things to exist in the absence of other forms of life. These small chlorophyll-containing cells could live very readily without us, and probably did for billions of years or more. Our lives today, however, and the lives of all other animals as well are dependent either directly or indirectly on the carbohydrates produced by green plants from solar energy, and the small green cells of the algae and the protozoa form the essential first link in the long food chain in which we, too, are interlocked.

Some of the protozoa that do not themselves possess chlorophyll have domesticated certain algae that live harmoniously inside them, presumably enjoying the protection of the larger animal and providing their host, in turn, with some of the rewards of their photosynthesis. This may prove hazardous for the algae, however, for certain "ungrateful" ciliates such as *Paramecium bursaria* will, in time of famine, digest their photosynthesizing companions. An interesting unsolved problem in protozoology is how, for example, *Stentor polymorphus,* when it ingests certain useful algae, does not devour them as it would any other food particle but releases them unharmed into its cytoplasm, or how *Paramecium bursaria* can suddenly "decide" to digest cells that were previously residing safely in its interior.

PHAGOCYTOSIS

Phagocytosis, although it may seem more advanced and complicated than the other eating processes, is actually the one that is the most thoroughly understood. This is the method by which solid particles are taken into the cell and, though it looks very different when an amoeba does it than when a paramecium does, the basic mechanism is the same. Protozoa have no mouths, properly speaking, in the sense of a permanent opening into the body, such as we have. The food must always pass in through the cell membrane. This is accomplished by the formation of a vacuole, a little balloon, or sac, pinched off from the cell membrane. In this may be enclosed a clump of bacteria, an algal cell, all or part of a fellow protozoan, or even a rotifer, which is a small many-celled animal. This vacuole then drifts throughout the body; in the amoeba it moves at random while the amoeba moves, but in certain of the ciliates it appears to follow a quite definite course.

At the beginning of the digestive journey, the victim is often still alive, and sometimes a thrashing rotifer or other seemingly indignant captive breaks through the membrane of the vacuole, but this fragile structure possesses remarkable abilities to repair itself. In the course of its travels, the contents of the vacuole are digested. Recent electron-microscope studies indicate that digestion takes place as a result of enzymes that are supplied to the food

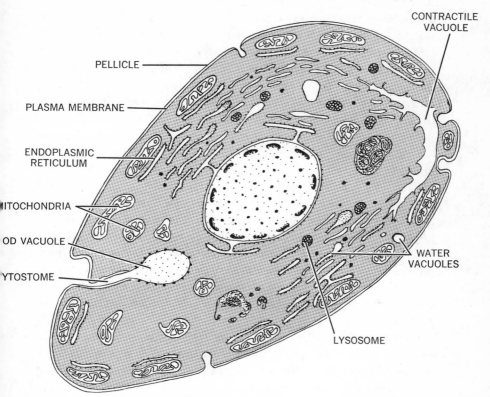

FIGURE 13   Diagram of the "digestive system" of the ciliate tetrahymena. (50 microns.)

particles from tiny sacs called lysosomes, present in the body of the cell, which fuse with the membrane of the vacuole and empty their contents inside it. Lysosomes have also been seen in some cells of our own bodies. They are sometimes called "suicide bags" because if they burst within a cell they can destroy it, actually causing it to digest itself.

The electron microscope shows that bits of food are often parceled off in smaller vacuoles and sent throughout the cell. The nourishing part of the meal diffuses through the membranes of these vacuoles, in a manner analogous to that by which the nourish-

33

ing parts of what we eat diffuse through the cells of the intestinal wall into the blood stream. The indigestible remainder is discarded, still in its vacuole, through the cell membrane. Some of the protozoa have a special opening, the anal pore (or cytopyge), through which the vacuole, greatly reduced in size and usually squeezed dry, is passed out, while others simply extrude the debris through the nearest available point in the cell membrane.

With the aid of the electron microscope, cytologists are discovering that another process, pinocytosis, also appears to play an important role in cell physiology. Pinocytosis simply means "cell drinking" as compared to "cell eating," or phagocytosis, and the chief differences are first, that only dissolved materials are taken in by the drinking process, and second, that the vacuoles formed are much tinier than in phagocytosis. Pinocytosis is being studied with special interest because it provides an answer to the puzzling question of how very large molecules, such as proteins, can cross the cell membrane. Pinocytosis has now been observed, quite unexpectedly, in many of the cells of our own bodies as well as in the protozoa.

Clearly visible within most of the fresh-water protozoa are the contractile vacuoles which pulsate rhythmically, serving as bilge pumps to expel excess liquids and, at the same time, undoubtedly, some dissolved waste materials. A contractile vacuole may have a fixed location, always pumping out through the same pore in the coat, or pellicle, as it does in euglena or paramecium, or it may move in the cell and simply expel its contents through any spot on the cell membrane. Its chief function, however, is not the excretion of wastes but the pumping off of excess water taken into the cell during phagocytosis or pinocytosis or by simple leakage through the outer membrane. Protozoa that live in salt water, even the close relatives of fresh-water forms, often do not have contractile vacuoles; the medium in which they live is just about the same density as their own protoplasm. Therefore, maintaining their own pressure against that of their environment is not a constant problem as it is among the fresh-water dwellers.

Food can be stored in the protozoa, in the form of starchlike compounds or fats in the plantlike protozoa, or sugar-like compounds in the other species. As in our own body cells, some energy

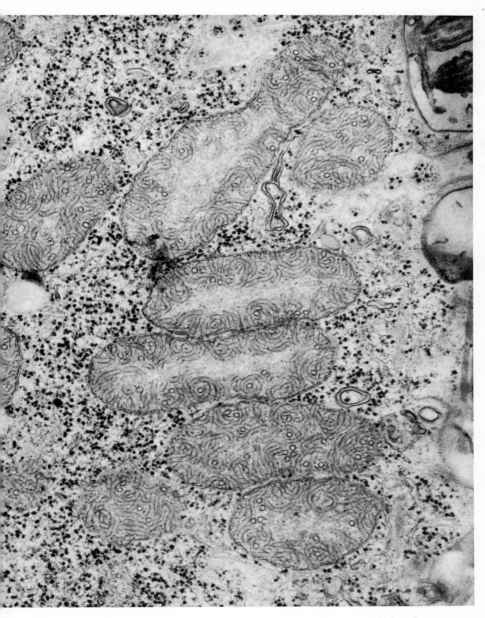

FIGURE 14   Mitochondria from paramecium, showing the microtubules characteristic of protozooan mitochondria. This type of mitochondrion is also found in insect flight muscles and in the adrenal glands. (Magnification 33,500 ×.) Micrograph courtesy G. Millonig and K. R. Porter.

35

is available for ready use in the form of a compound known as ATP (adenosine triphosphate). As the name of this chemical implies, it contains three phosphorous groups; these are attached to the body of the molecule by high-energy bonds. When the phosphate group is snapped off—there are special enzymes for this purpose—quick energy is made available, just as in lighting a match. Scientists have recently discovered that the chief factories for the production of ATP are the mitochondria. With the light microscope, if you use special staining techniques, you can see mitochondria as small potato-shaped bodies, which sometimes fuse to form long tubes. In the living animal, they are often found clustered around particularly hard-working structures in the cell, such as the contractile vacuoles.

The electron microscope shows that the inside of the mitochondrial bodies are filled with convolutions of membranes, and biochemists have proved that the enzymes that convert the breakdown products of sugar or starch to the quick-energy ATP are actually embedded in these membranes. Bacteria and blue-green algae— generally considered more primitive than the protozoa—lack mitochondria, but in these cells the mitochondrial enzymes appear to be embedded in the outer cell membrane. This is an interesting example of evolution—the development of a specialized system to provide more membrane surface for a highly useful adaptive function.

HUNTING AND FORAGING

Although the methods of handling food are remarkably similar in all cells, the ways in which protozoa trap their prey are highly varied and idiosyncratic. *Amoeba proteus,* for instance, may simply run over small particles, mopping them up, but can be more frequently observed to extend pseudopodal arms around its prey, engulfing it. The pseudopods then fuse and the cell membrane enveloping the organism is pinched off from the cell surface with the captured organism trapped inside. These arms are very strong. A large ciliate, such as paramecium, may be cut in two by the pincer action of two converging pseudopods.

This same method of phagocytosis is also seen in the digestive cells of certain many-celled animals, such as flatworms and jellyfish. These cells simply grab food particles as they pass by and process them to feed the other tissues. It is also the means by which our own white blood cells, the phagocytes, deal with bacteria and other foreign organisms they encounter in the blood stream.

The soil-dwelling amoebas have developed a slightly different method. Characteristically, they have thick, warty-looking (verrucosid) pellicles. Their favorite food consists of the small, shelled amoebas called testacea, and tiny flagellates that also dwell in films or droplets of water caught between soil particles. They can apparently sense either of these animals at some distance—at least an amoeba-length away—and can be seen to extend a long, purposeful pseudopod directly toward their intended prey. When the pseudopod touches the testacean, it adheres to it; meanwhile the victim has futilely retreated into its shell. Then a peculiar phenomenon occurs. The verrucosid amoeba apparently digests a hole in its own thick pellicle large enough to accommodate its prey, shell

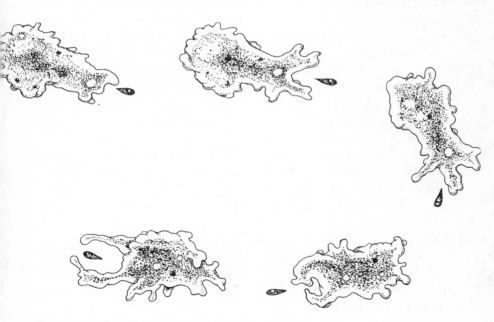

FIGURE 15   Amoeba in pursuit of a small flagellate.

and all. The protoplasm at the end of the pseudopod softens to sol (see page 14), drawing in the testacean. Then the pseudopod becomes a suction tube as the protoplasm withdraws along it and the tube constricts behind it. At the bottom of the tube the prey is enveloped in a swatch of cell membrane and the process of digestion begins. It takes about fifteen minutes for the amoeba to sense its prey, capture it, suck it in, and package it for consumption.

In the case of the small flagellates, the amoeba extends a feeding pseudopod and grabs it by one flagellum. Then it simply holds tight, usually for about four or five minutes, until the flagellate stops struggling. Then the amoeba begins to suck on the flagellum and continues until it draws the entire little animal inside.

THE OMNIVOROUS FLAGELLATES

The green flagellate euglena does not engulf solid food particles at all, relying on photosynthesis and saprozoic nutrition alone, but its close relative peranema is distinctly a phagocyte and, indeed, one of its favorite foods is cousin euglena. When the tip of its extended flagellum comes in contact with euglena, the flagellum beats actively and the whole body of the animal contracts and elongates several times. Then peranema stretches open its cytostome, which is on the underside of its body. Next to the cytostome are two sharp, transparent rods, or trichites. Each of these rods is composed of a bundle of about one hundred parallel, hexagonally packed, tubular fibers. The two rods are joined at their base by a bridge so that they move as a unit. Peranema first seizes euglena by these trichites, holding it as if on a two-pronged fork, gulps in as much as possible of the animal, withdraws its trichites, and then jabs them in again, getting a fresh purchase by which still more euglena can be engulfed. The whole flagellate can be swallowed, a gulp at a time, in as little as two minutes, no longer than fifteen. Peranema can literally open itself up like a large sack to stuff inside a euglena almost as large as itself. When it encounters a euglena too large to swallow—that is to say, bigger than itself—it bores a hole in its side with one of its trichites and draws out the protoplasm.

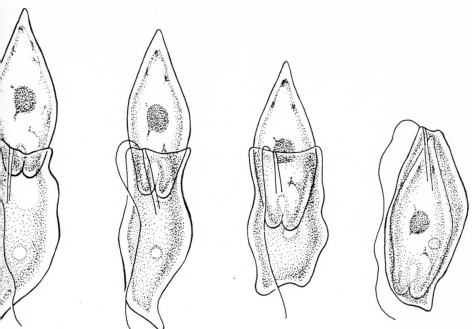

FIGURE 16 Peranema engulfing euglena.

EATING AMONG CILIATES

The ciliates are noted for the variety and ingenuity of their mouth structures; and, indeed, this is a chief basis for classification among this subphylum. In paramecium, for instance, a deep channel, the oral groove, leads along the underpart of the animal's body down into the cytostome. When paramecium is swimming slowly or browsing, the cilia around its oral groove beat more strongly than those of the rest of the body and so they collect particles from a distance, which flow down the funnel-like oral groove. Some particles are rejected by the beating of the cilia in the vestibulum, the outer entrance to the groove, while others, accepted, rush down in, pressing against the delicate membrane of

39

the cytostome until it balloons in and seals off. Then almost immediately another vacuole begins to form. A single paramecium may devour as many as five thousand bacteria a day in this way. The food vacuoles move through the cytoplasm, and often many may be seen in the same cell with their contents in various stages of digestion. Within the kidney-shaped colpoda, for instance, a common soil-dweller much smaller than paramecium—in fact, usually less than one hundred microns in diameter—some two hundred food vacuoles have been counted in one cell. The early microscopists considered these vacuoles to be little stomachs, just like our own, only smaller, and called these ciliates, quite understandably, the "polygastrica."

Didinium, also a ciliate, has a quite different method of feeding than paramecium and other polygastrica. Instead of living on bacteria ingested in compact masses, it eats large whole animals. It lives almost exclusively on paramecia. Didinium is barrel-shaped with a girdle of cilia that whirl it rapidly through the water. Its cytostome is at its anterior end, on the middle of the barrel lid, and on top of the cytostome, which is very small when closed, is a small sharp horn, or proboscis. Lining and supporting the area beneath the cytostome, like the whalebones in an old-fashioned corset, is a circle of trichites. Didinium attacks the paramecium proboscis first, holds it momentarily by the lip of its cytostome, and then the whole structure opens wide to engulf the entire prey. Didinium can swallow a paramecium twice as large as itself, stretching its whole body around its captive in one giant food vacuole, and it can repeat this gluttonous performance with remarkable frequency devouring as many as a dozen a day. Not only must paramecium satisfy the voracious appetite of its fellow ciliate, but it must also, scientists have shown recently, provide the wherewithal for its own digestion. Didinium apparently lacks the digestive enzyme dipeptidase and depends on the luckless paramecium to supply it.

Obviously, a pack of didinia can rapidly exhaust even a large colony of paramecia. When this occurs, each didinium simply forms a dry shell, or cyst, around itself and waits in a state of suspended animation until the crop of paramecia renews itself. Then didinia excyst, emerging to begin the slaughter anew. Thus

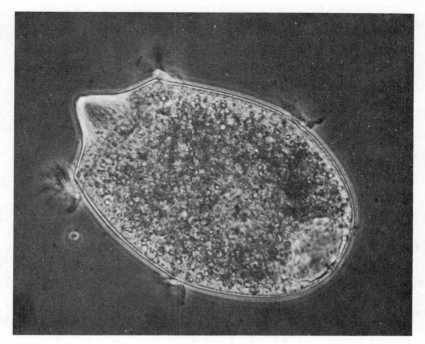

FIGURE 17   Didinium, showing the small sharp proboscis with which it jabs paramecium. (80–200 microns.) Photograph by Eric Gravé.

didinia and paramecia exist in an inverse proportion to each other —a ratio that can be plotted quite precisely by two curves—each balancing the other down through the eons of time.

THE SUCTORIA

The suctorians have approached the nutrition problem in quite a different way. They spend most of their life anchored in one spot, rigid and motionless, but with deadly tentacles extended like spikes. The best-studied of the suctorians is perhaps *Tokophrya infusionum*. When seen from the side, tokophrya is a little pear-shaped animal anchored to the substrate by a short stalk and flat disc. Protruding from the top of its body are two clusters of what

41

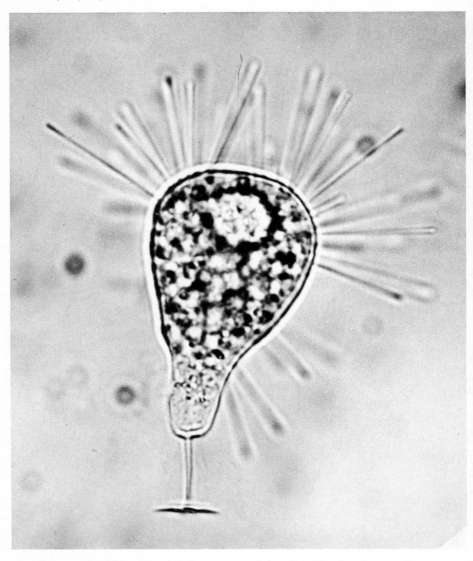

FIGURE 18 *Tokophrya infusionum,* seen from the side, showing attachment disc, stalk, and tentacles. (60 microns.) Photograph courtesy Maria Rudzinska.

appear, under the light microscope, to be long slender rods, fragile as fine-spun glass. The animal is small, usually less than fifty microns high, and the tentacles, which may be longer than the body itself, are less than a single micron in diameter, except at the tip, where they round out into a knob. One of tokophrya's favorite foods is its fellow ciliate tetrahymena, which somewhat resembles the more familiar paramecium in shape, although it is much smaller. When a luckless tetrahymena brushes across an out-stretched tentacle of tokophrya, it stops, apparently stunned, and tentacle and prey become firmly attached. Tokophrya is smaller than tetrahymena; but, despite its small size and the seeming fragility of the rodlike tentacle, the larger ciliate can almost never free itself, and often its attempts to break loose succeed only in its becoming fastened to additional tentacles of the captor. Once at-tached, the tentacle broadens and shortens; and, as painstaking light microscopy studies have shown, a stream of tiny granules moves up the rod to the tetrahymena. The struggling animal be-comes immobilized. If the captive is rescued from the tentacles at this stage, it will remain motionless from several minutes to several hours, depending on the duration of its exposure, but in most cases it regains its normal mobility.

A larger ciliate, such as paramecium, may frequently break free from a single captor. Often, however, tokophrya is found in clusters, the spikes almost touching, like the barbs in a roll of barbed wire, and the paramecium that wanders into this no man's land rarely escapes. It touches one tentacle, pulls loose, seemingly stunned, and staggers into another, until it is entirely immobilized. Sometimes a single tokophrya will fasten into paramecium with such a firm grip that it is pulled loose from the substrate and swept off, still attached to the relatively giant animal. In such cases paramecium soon slows down and finally stops; then toko-phrya reanchors itself and begins to feed. Clearly the tentacles are the source of some toxin—a sort of protozoan curare—although it has not been isolated or identified. Curiously, the tentacles adhere only to ciliates; the rhizopods and flagellates pass without danger.

Once the prey is attached to the tips of one or more tentacles of one or more tokophryas, the poison-producing barbs become sipping straws. One can see a flow in the other direction, from

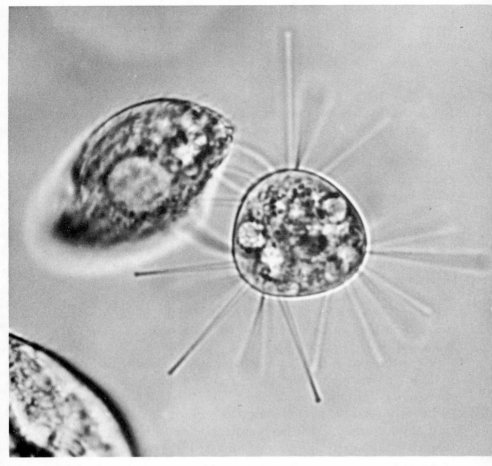

FIGURE 19   Tokophrya feeding on tetrahymena. Photograph courtesy Maria Rudzinska.

tetrahymena to tokophrya. The prey remains alive during the feeding process; the continued pulsing of its contractile vacuole is readily visible. In fact, as soon as the victim dies, tokophrya discards it as unfit for consumption. The whole process takes about twenty minutes.

It is presumed that the tentacles exert a sucking action (hence,

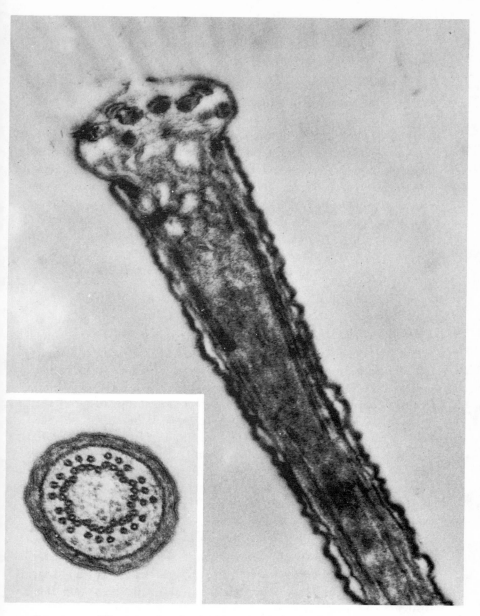

FIGURE 20 The tentacle of tokophrya (above), as shown by the electron microscope. The dark bodies in the knob are the missiles. One of them has penetrated the outer membrane of the tentacle and you can see its pointed end protruding. The cross section (inset) shows the bundles of microtubules, the plasma membrane, and the pellicle. (Tentacle magnification 38,400 ×; cross section magnification 70,620 ×.) Photographs courtesy Maria Rudzinska.

45

the order to which tokophrya belongs is called the suctoria), but just how the suction is produced is not clear. Some have suggested it is a result of rhythmic, wavelike contractions of the tentacles, and some believe increased pressure within the body of the victims may drive its contents down the tubes. Others have suggested that increased action of the contractile vacuoles may turn tokophrya into a tiny, lethal suction pump, but none of these explanations seems adequate. In any case, the feeding mechanism is extraordinarily efficient. All of the fifty or sixty tentacles can feed simultaneously, should nature or a generous protozoologist provide such bounty. And tokophrya is a glutton. It will feed continuously for forty-eight hours or more, and under such conditions will grow to 120 times its normal volume.

The electron microscope has revealed that the tentacles, despite their small size, have a very complicated structure. Each one is made up of two concentric cylinders. The outer wall of the outer cylinder is covered with a delicate plasma membrane—the same sort of membrane that covers the surfaces of all cells, including our own. In addition, outside this membrane is a more rigid and thicker coat, the pellicle. The plasma membrane covers the entire tentacle, including the lumpy surface of the knob; but the pellicle reaches, like a coat sleeve, only to the base of the knob.

The wall of the inner cylinder is composed of a double row of forty-nine—or, occasionally, fifty—microtubules arranged in loose bundles of seven tubules each, four on the inside, three on the outside. These tubules, which extend deep within the cell, are thought to be contractile and so responsible for the shortening and broadening of the tentacle during feeding. The inner tubule, usually only one fourth of a micron in diameter, expands greatly during feeding so that even relatively large morsels—mitochondria, for instance, which are one micron wide—can pass smoothly through.

The outer cylinder is the carrier for the tide of minute, dense ovate bodies, the bodies that in light microscopy can be seen flowing from predator to prey immediately following capture. These bodies are missile-shaped, rounded on one end, pointed on the other. These missiles apparently arise in the cytoplasm of tokophrya and travel up the outer tubule of the tentacle to the knob, where they pierce the naked plasma membrane with their pointed ends.

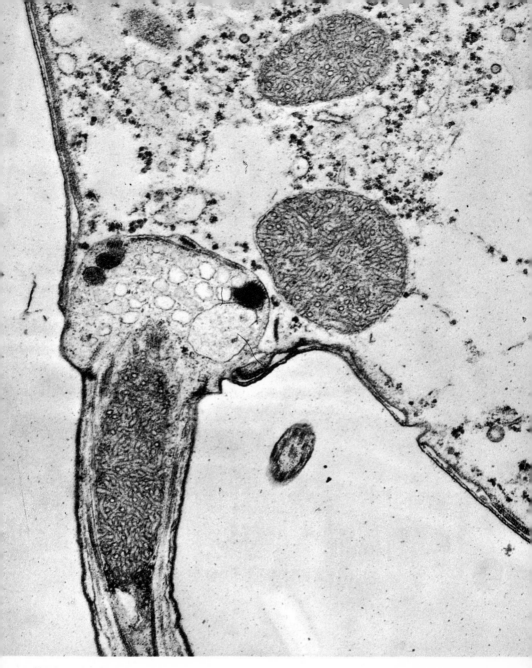

FIGURE 21 The knob of tokophrya is embedded in the body of its prey and the pellicles of the two cells have become continuous. Three mitochondria of tetrahymena can be seen, one of which is moving down the tentacle. (Magnification 30,000 ×.) Photograph courtesy Maria Rudzinska.

These ends, which protrude from the knob, look fuzzy under the electron microscope. This fuzziness may well mean that they are sticky, and probably it is the protruding missile ends that cling to the cilia of the prey and hold it fast. They probably also secrete the paralyzing poison. The missiles also seem to be the carriers of a number of enzymes, which perhaps serve to dissolve the pellicle of the captured ciliate and also to make its cytoplasmic contents less viscous, facilitating their smooth flow down the tentacles. It is not known how the missiles work, but apparently they discharge their contents from the tip of the knob; their dense, characteristic forms have never been found outside tokophrya or within the body of the prey.

Electron micrographs taken of the knobby end of a feeding tentacle reveal another unexpected phenomenon. The knob is apparently driven like a fist into the soft body of tetrahymena; once this occurs, in a matter of microseconds, the pellicle, or outer sleeve, of the tentacle becomes continuous with the pellicle of the captured ciliate and the struggling animal cannot free itself from this seemingly fragile rod. Also, of course, this strange union results in a perfect seal, so none of the contents of the captive can leak out or spill. In less than a second the two become a single organism and remain so until the captive is dead.

DEFINED MEDIA

For years, scientists have attempted to grow protozoa on media in which every ingredient is chemically defined. This is not too difficult for the phytoflagellates, since their chief energy source is photosynthesis, but it has proved very difficult for the other organisms. First, of course, the culture must be completely free of bacteria or any other living things, except the one species under study, which means working under the most meticulous conditions with not only all equipment but the air itself maintained absolutely sterile. Secondly, the scientist must know every single material that is present; a number of protozoologists have thought from time to time that they have worked out the precise dietary details for a culture to find out, for example, that the medium was contaminated with a speck of lint from the test tube's sterile cotton plug, which supplied some unknown but absolutely essential carbohydrate or with some undefined mineral leached out from the laboratory glassware. The first nonphotosynthesizer for which the requirements have been exactly defined is the ciliate tetrahymena, which has been found to need some forty-three separate ingredients, including seventeen amino acids, nine mineral salts, and a few miscellaneous items. Although tetrahymena feeds by phagocytosis, nine-tenths of this liquid diet is absorbed saprozoically, illustrating once again the remarkable adaptability of these marvelous animals.

These studies on tetrahymena have made clear that its nutritional needs—and of course, by implication, its physiological processes—differ little from those of the cells of the chicken, the rat, or indeed of our own tissues. In fact, our own cells require exactly the same amino acids and vitamins as tetrahymena. As a consequence of these unexpected findings, tetrahymena and other protozoa as well are being used increasingly for biological assays of nutrients—as is euglena for B_{12}—an employment for the one-celled animals that would surprise the classical protozoologists. Some scientists regard them also as almost ideal model systems for fundamental research on problems of cell biochemistry.

CHAPTER IV

REPLICATION, BIRTH, AND DEATH

Man produces a modest number of offspring compared to almost all other animals, but the individuals tend to survive. Any single paramecium, on the other hand, has little reason to anticipate a long life, but each single one can replicate itself so rapidly as to touch off a chain reaction—two, four, eight, sixteen, thirty-two, sixty-four, one hundred twenty-eight—that soon reaches an astonishing total. In fact, one patient protozoologist once calculated that if all the progeny of any single paramecium survived (assuming a division rate of once a day), in 113 days there would be a mass of paramecia equal to the volume of the earth. Some species, in fact, can divide as often as five times a day, which would greatly hasten this process. It is this high rate of reproduction that permits the survival of the one-celled animals and also, in all probability, the development of the many species and varieties of protozoa.

Protozoa most commonly multiply by splitting in two, the same general process by which our own bodies develop from one fertilized egg into the billions of cells that make up the adult human body. In the studies of division among somatic cells—the cells which compose the tissues of multicelled plants and animals—attention has been focused on the events taking place in the nucleus. The nucleus, which is generally regarded as the cell's "control center," is the storage site for the hereditary material, a group of enormously complex molecules known as DNA (deoxyribonucleic acid). DNA, on chemical analysis, has proved to be a chain of relatively small submolecules, many thousands of them; these submolecules form the genetic code, actually spelling out the hereditary messages in a manner analagous to the way in which the letters of our alphabet spell out the sentence you are now reading. DNA serves two functions: First, by faithfully replicating itself, it preserves the hereditary information that is passed on from cell to cell during cell division. Second, with the help of another similar chemical, RNA (ribonucleic acid), it controls the moment-by-moment production of proteins. Proteins are the basis for many cellular structures; they are also the material from which enzymes are made—the chemicals on which all cellular function depends. The gene, the unit of heredity, has now been redefined: It is the section of a DNA molecule that dictates the structure of a particular protein.

Before cell division, the DNA duplicates itself, with one copy for each daughter cell. Then, at the time of division, or mitosis, the DNA, which is usually diffused throughout the nucleus, condenses and becomes visible with its associated proteins as sausage-shaped chromosomes, each paired with a partner that is its exact twin, the end result of the copying procedure. In the course of mitosis, these pairs of chromosomes separate, each migrating to an opposite pole. These poles are marked by two small cylindrical bodies, the centrioles, from which radiate the spindle fibers that appear to pull the chromosomes apart. When the two groups of chromosomes are separated and at opposite ends of the cell, each enclosed in its own nuclear membrane, the outer cell membrane constricts around them to form two new daughter cells, each of which contains approximately half of the previously existing cytoplasm. Among somatic cells only relatively simple, or undifferentiated, cells divide. The highly specialized ones, such as the nerve cells, cease dividing very early in the life of the organism.

In the protozoa, the situation is more complicated. In general, protozoan cells are much larger than somatic cells. They often possess more than one nucleus, apparently needing more than one "control center," to command so vast and complicated a field. In the case of the ciliates, two different types of nuclei have evolved one of which, the micronucleus, seems to serve the function of preserving the hereditary material, the other of which, the macronucleus, directs the cellular activities. They also possess many more different kinds of structures than somatic cells. It was long believed that replication of cellular structures was controlled entirely by the nucleus. It has now been discovered that many of the cytoplasmic structures of the protozoa are self-replicating, and indeed that they contain their own DNA. Much of the work on "extrachromosomal inheritance," as it is known, has been done with the protozoa, and now these findings, once thought peculiar to the protozoan and a few other "simple" cells, are beginning to shake some of the foundations of classical cytogenetics.

CELL DIVISION IN FLAGELLATES

The four major subdivisions of the protozoa—the flagellates, rhizopods, ciliates, and sporozoa—have characteristically different methods of cell division. The flagellates usually split in two longitudinally, from front to back, each half forming a mirror-image of the other. In euglena, for instance, the nucleus, which generally lies near the center of the body, moves forward until it is close to the reservoir. The cell begins to divide at its anterior end, losing the free flagellum in the process. The bases of the two flagella divide, and new flagella form very rapidly from these bases; at one point in division there are four short flagella all beating within the same, shared reservoir. The chromosomes become visible as the DNA condenses, and the nucleus then begins to pull apart forming a dumbbell-shaped structure (the nuclear membrane does not disappear as it does during mitosis in somatic cells). As the cell continues to divide, new structures are formed, and by the time the two nuclei are completely separated each cell has a completely new anterior, with its own reservoir, rudimentary flagellum, free flagellum, and other organelles. The beating of the new flagella serve to pull the still dividing cells apart.

The bases from which these flagella develop so rapidly have been known by various names—blepharoplasts, or basal granules, probably being the most frequent—while those from which cilia developed came to be known as kinetosomes. The electron microscope has shown that all these basal bodies share the same structure, and the modern tendency has been to call them all kinetosomes, or "moving bodies." It has also been found that the centrioles of

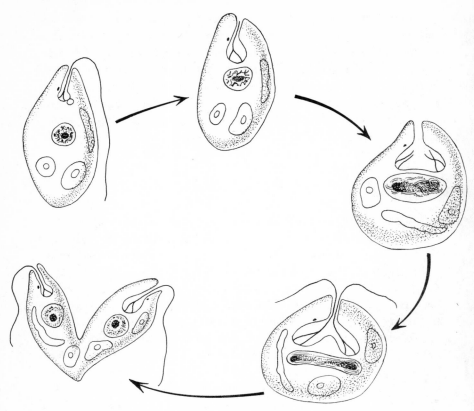

FIGURE 22 Euglena and other flagellates divide longitudinally, each new cell forming a mirror image of the other.

cells, including the cells in our own bodies, have the same structure as the kinetosomes. In some cells the centriole serves a double function; in the human sperm, for instance, a centriole first helps to organize the spindle, and then, once cell division is complete, grows a flagellum, the sperm tail.

The new euglena cells are "young"; they are smaller than the parent, more vigorous, and consume more energy, but they are in no sense infants. In fact, among most of the protozoa, the entire development is completed before reproduction, and the new cells begin life as formed adults.

In the case of the dinoflagellates, cell division is often oblique rather than strictly longitudinal. Among those that are armored with heavy cellulose plates, such as gonyaulax, each daughter cell receives approximately half of the cellulose overcoat and then provides for itself whatever parts of the often quite elaborate coverings are missing. Many of the flagellates go through a "palmella" stage, in which they lose their flagella and secrete a sort of protective jelly around themselves. Sometimes the cells divide in the palmella stage, remaining stuck together by their gelatinous exteriors. Palmelloid dinoflagellates may divide within the cellulose wall of the parent cell, and then break it open as if it were a cyst, or they may remain indefinitely in their jelly bed, some of them, in fact, spending a major portion of their life cycle in this form.

DIVISION AMONG RHIZOPODS

Because of their protean form, the amoebas cannot really be said to divide along any axis. *Amoeba proteus* prepares for division by withdrawing its larger pseudopods and assuming a shape resembling a chestnut bur, studded with very short pseudopods. During this process it becomes steadily less responsive to stimulation—such as the presence of an edible flagellate. The nucleus, ordinarily visible, disappears, but special staining techniques can reveal the chromosomes, in some species as many as five or six hundred of them in a single cell. These chromosomes, which are very small, arrange themselves on the spindle and separate. As nuclear division takes place, the cell elongates, a constriction appears in the middle of it, and then the two daughter cells begin to tug apart, each crawling in opposite directions on its own fully formed pseudopods.

Problems arise in binary fission among the testacea, the rhizopods that have special outer coverings, or shells. Usually one daughter retains the shell while the other must provide itself with a new one. Arcella, for instance, is a rhizopod covered with a translucent, bowl-shaped shell of its own making. At the time of division, arcella protrudes a portion of cytoplasm, like a large pseudopod, out from under this inverted bowl. Then this pseudopod begins to secrete a membrane about itself, which soon thickens and becomes a new cell. During this process, the protoplasmic connections between the two cells constrict, but not until the ejected daughter has completed its new shell do the two pull entirely apart.

Other shelled forms appear to display less concern about the fate of the dispossessed partner, and in many cases one is simply ejected and swims around naked until it settles down to form a new shell, or lorica.

Some of the giant amoebas reproduce by multiple fission. These cells, which may reach a size of three millimeters or more, usually contain a number of nuclei. These rhizopods simply break up into two or three—sometimes more—pieces, often of disparate size, but each containing a share of nuclei, contractile vacuoles, and other essential equipment. This rather inelegant method of reproduction is known as "plasmotomy."

DIVISION AMONG CILIATES

Unlike the flagellates, which split longitudinally into mirror-images, the ciliates characteristically divide transversely, with one animal forming on top of the other; the anterior cell produced by this sort of division is known as the "proter"; the bottom cell, the "opisthe." In paramecium, the first sign of cell division is an elongation of the body of the parent animal. As this occurs, new kinetosomes form along the longitudinal rows, preparing for the formation of additional cilia. The oral groove disappears. The kinetosomes of the oral cilia duplicate themselves, form a new group of kinetosomes, which migrate down the elongated cell body, and organize themselves to form the oral ciliature of the opisthe. Some thirty minutes after the whole process starts, the oral grooves then re-form and the two new paramecia separate.

Paramecium, like all the other ciliates, has two types of nuclei, a micronucleus (some species have two or three or four) and a macronucleus. The micronucleus apparently stores the hereditary information for safekeeping but does not participate in the ongoing activities of the cell. At the time of cell division, the mi-

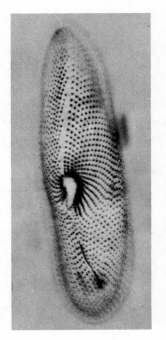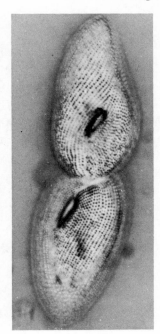

FIGURE 23 Most ciliates divide transversely, one above the other. These specimens of paramecia have been treated by a silver staining technique which reveals the ciliary pattern. Photograph reproduced from *The Scientific Endeavor*, T. M. Sonneborn, published by The Rockefeller University Press, 1965, Fig. 26, page 231.

cronucleus divides mitotically, so that each new cell receives a faithful copy of the heredity blueprint. The macronucleus is believed to contain the same information as the micronucleus, but repeated over and over again. It is the macronucleus that governs the day-to-day life of the cell; paramecia from which the micronucleus has been removed can survive indefinitely, and in fact some natural strains do not possess micronuclei at all. The macronucleus does not undergo mitotic division but simply elongates as the cell does and is finally pulled apart in two equal chunks, one going to each new paramecium.

STENTOR

In stentor, division is the same in principle but a little different in some particulars. Stentor has prominent bands of cilia running longitudinally down its trumpet-shaped body; a heavy crown, or wreath, of beating cilia at its anterior end, around the gullet; and a readily visible macronucleus, which consists of a string of nodes distributed evenly through the body. These nodes are all enclosed in a single, though usually invisible, nuclear membrane. The tiny micronuclei, of which there are several, lie slightly off to one side of the macronuclear chain.

The first sign that stentor is going to divide is a crescent-shaped interruption in the longitudinal ciliary bands at just about the midsection of the animal. Then an extra row of heavy cilia becomes visible curving up the side of the animal; these cilia begin to beat as soon as they appear, always in synchrony with the oral ciliature of the proter. If these new oral cilia are removed at this point, all cell division ceases. Obviously cell division, at least in the ciliates, involves an exchange of information between cytoplasm and nucleus.

Then the nodes of the macronucleus coalesce into a single central mass. The micronuclei migrate to the opposite ends of the cell; they divide mitotically but not, apparently, until cell division is complete. Then two fission lines appear, constricting the cell at the midline. As the cell divides, the crescent of new cilia moves into its appropriate position around the anterior end of the opisthe. When the ciliary wreath is complete and in place, the gullet begins to form. The macronucleus then elongates and is pulled in two; it re-forms into nodules, with each new cell receiving as many nodules as were possessed by the parent cell, but each one only half as large. Finally, the two cells are held together by only a thread of cytoplasm, attaching the tail pole of one to the anterior of the other. At this point the two daughters, which have synchronized their swimming motions during the six hours or more it takes for cell division, give a little twist in opposite directions, and part company.

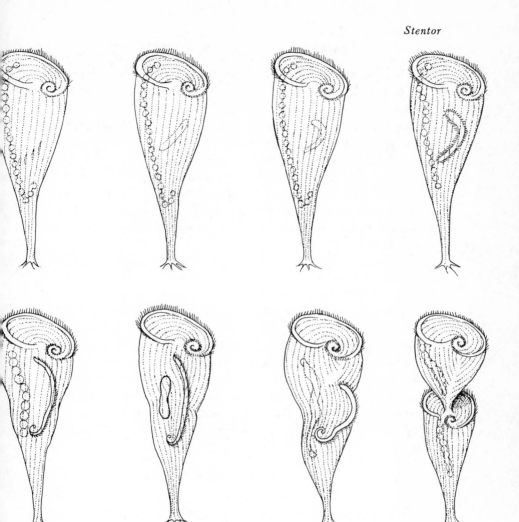

FIGURE 24   Stages in the division of stentor. (1–2 millimeters extended.)
After Tartar.

In stentor the macronuclear material of the parent cell seems to
be divided more or less equally among the daughters, leaving each
with a series of nodes which are at first half-sized (compared to
those of the parent) and which increase in volume as the new cells
grow to full size. In euplotes the situation is somewhat different.

The macronucleus of euplotes is very prominent, resembling a large sausage bent into the shape of the letter c; it is estimated to contain two hundred times as much DNA as the small micronucleus that lies beside it. Prior to cell division, clear strips appear at each end of the macronucleus; these are known as the reorganization bands. These bands move slowly toward one another, passing through the fibers of the macronucleus like combs. What happens—and you can almost see it—is that the genetic material doubles as these bands pass through. When the two bands meet, the macronucleus condenses into a thick rod. Meanwhile the micronucleus has divided mitotically, revealing its eight, quite prominent, chromosomes.

As with stentor, the ciliature around the mouth of the parent is retained for the proter, and a new ciliary apparatus is formed for the opisthe. New ventral cirri—fifteen to twenty, depending on the species—are made for both new cells. The new ones often appear before the old ones are discarded, so that for a short period the undivided organism possesses three sets. No one has been able to tell what happens to the old cirri. A cirrus simply stops its movement, becomes thinner and shorter, and then vanishes—all in about two minutes. As with paramecium, the new oral ciliature of euplotes, which is very large and complex, appears to have its origins in the kinetosome system of the parent's ciliature.

VORTICELLA

Not all ciliates divide transversely. Vorticella, shaped like a bell, presents special problems because of the firm attachment of the parent cell to its substratum (a word which means simply "foundation," whether the ground, a speck of debris, the leaf of a plant, the gill of a clam, or even another protozoan). In preparation for cell division, vorticella contracts into a sphere, drawing in its oral ciliature, and then broadens. A longitudinal fissure appears at the top of the cell and travels downward, but not quite through the center, as in the mirror-image division of flagellates. One cell retains the old feeding apparatus and the stalk, as well as the bulk of the cytoplasm, and more than half of the macronucleus. The

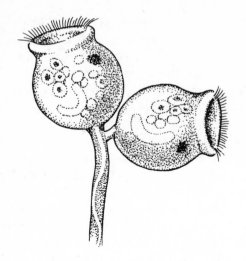

FIGURE 25   Vorticella budding. The daughter cell, now full grown, will soon break away from the parent stalk. (50 microns.)

daughter cell develops a full set of oral ciliature while still attached to the parent, as well as all its necessary internal organelles. When it is ready to leave—the whole process takes only twenty to thirty minutes—the parent snaps back and forth on its highly contractile stalk to shake the daughter loose, while the daughter cell rotates, beating its cilia, to twist off the last remaining cytoplasmic tie. Finally, at the last moment, the daughter contracts itself into a little cylinder, sprouts a circlet of cilia around the base of the cell, like tail feathers, and swims away. Not long after, the telotroch, as this swimming form is called, will settle down, grow a stalk of its own, lose its circlet of cilia and become a vorticella just like its parent.

THE SUCTORIANS

The suctorians, which are also all stalked ciliates, solve their problem in somewhat the same way as vorticella. Each parent cell produces one or more buds at a time; these buds, in contrast to most of the "newborn" protozoa, are not complete animals but are immature, or larval, and do not grow into adulthood until they settle down and grow their own stalk.

Most protozoa, through ceaseless cell division, achieve a kind of impersonal immortality, but the suctorians share with man and other many-celled animals an identifiable life span that ends with aging and eventual death. This process has been observed in particular in tokophrya which, as it grows older, undergoes a series of changes which have interesting correlations with those seen in aging cells in our own bodies.

Young tokophrya is almost perfectly spherical, about twenty-five microns in diameter, and has an average of ten strong, firm tentacles. For the first twelve hours or so of its sessile life it feeds actively, growing more tentacles, but is unable to reproduce. Its macronucleus is a compact dark mass which, in the electron microscope, can be seen to be made up of a number of small, dense granules, the chromatin bodies. The most prominent of its cytoplasmic organelles are its numerous mitochondria scattered uniformly throughout the cytoplasm.

By the end of its first day tokophrya is a young adult, at the height of its powers. During this period of vigorous maturity it possesses an average of fifty functioning tentacles and is able to produce as many as twelve embryos during a twenty-four-hour period, each supplied with a micronucleus, produced by mitosis, and with a pinched-off segment of macronucleus. During this period its body, now about forty microns across, assumes a more definite pearlike shape.

After four or five days of this peak activity, the cell begins to age. It grows slightly larger and more irregular in outline. The number of tentacles dwindles and those remaining grow weaker; tetrahymena, its favorite prey, can now quite easily break away, and even when a tentacle is firmly attached the aging cell takes as long as two hours to feed. Reproductive powers wane until only one or two embryos are produced in a day. The number of mitochondria decreases, and those that remain tend to gather around the periphery of the cell. Cross sections of the mitochondria in aging cells show signs of breakdown in their internal structures, the enzyme-holding membranes. The cell begins to look bloated, filled with vacuoles that contain small granules which appear to be insoluble waste materials. Unlike almost all other animals, including one-celled animals, tokophrya lacks any means for disposing of

these insoluble wastes, and their accumulation may be directly related to aging. Grossly overfed tokophrya begin to show signs of old
age more rapidly than animals on a more Spartan diet, and it is
probably that the near-starvation diet of tokophrya in its natural
habitat may favor its long survival, as indeed does undereating in
many animals, including, sadly, man.

Certain cells in the mammalian body also grow old; these are
the cells of the central nervous system, purkinje and other nerve
cells, that cease dividing early in the life of the organism. The
most characteristic finding in these aging mammalian nerve cells
is the accumulation within the cell of what is known as "senility
pigment"—pigmented granules of lipid-like material. Lipid granules similar in appearance have been found in tokophrya. Both in
tokophrya and in human neurons, the pigment is never found in the
young cells; it begins to accumulate in the adult and is always
present in the old cells. Electron micrographs of aging tokophrya
show that these granules accumulate in the region of the mitochondria, and some of the photographs indicate strongly that they
actually arise from the mitochondria themselves, perhaps as these
bodies degenerate with age. On the evidence available, the cause
of death in tokophrya, and perhaps in nerve cells as well, may be
from autointoxication, caused by the animal's inability to eliminate
these waste products.

Another conspicuous feature of the aging cell is the degeneration
of the macronucleus, which grows much larger, reaching three or
four times the diameter of the macronucleus in the adolescent
animal and becoming irregular in shape. The chromatin bodies
increase in size and number and decrease in density. In some cases
the macronucleus begins to undergo divisions, and sometimes as
many as four macronuclei are found in one aging cell. Additional
nuclei are also found in aging purkinje cells, and it might be that
this phenomenon reflects in both cases an attempt by the cell to
repair its central control machinery in order to regenerate its ener-
gies and prolong its life.

Finally, unable to reproduce or eat, reduced to a mere fifteen
worn, weak tentacles, its mitochondria emptied, and its macronucleus in disarray, tokophrya dies. Its entire cycle of life and
death takes just ten days.

CHAPTER V

SEX: MEIOSIS AND MATING

When cells increase by simple division, all of the new cells are essentially identical twins. Any changes that occur take place by chance—the result, perhaps, of a mistake in "copying" the DNA—or by damage from natural or man-made radiation. These changes are seldom favorable; if you take one word out of a sentence and put in another, selected absolutely at random, there is very little chance that the sentence is going to make better sense than it did before. Consequently few of these "mistakes," or mutations, persist, and the general cellular population of identical twins carries on virtually unchanged over great periods of time, stable and conservative. This can have its advantages. Some of the protozoa—euglena and amoeba, for instance—are known to reproduce only asexually, by simple cell division, and their success by all biological standards cannot be questioned.

Genetic diversity, on the other hand, is also desirable. Without variety there could be no evolution, and the forces of natural selection would be offered no choices. Without constant variation there could not be the rapid bursts of evolution such as resulted in the dinosaur, not so very long ago, and in modern man. Genetic recombination—more simply known as sex—is the most reliable system for introducing variation, and almost all forms of living things, including bacteria and viruses, have worked out some system for achieving such recombinations. The protozoa have experimented with several.

SEX IN MAN

Sex in man involves two processes, meiosis and mating. Meiosis (from a Greek word meaning "to make smaller") is the process by which the chromosomes of either the primary spermatocyte in the male, or the primary oöcyte in the female, are reduced to half their number. In meiosis the forty-six chromosomes join in twenty-three pairs; these pairs represent the original chromosomes from the parents, one chromosome of each pair being "maternal" and the other "paternal." Then the pairs undergo a sort of fusion process, during which portions of the paired chromosomes may be exchanged. In this way if an individual has received a gene from the mother, located on one chromosome, and a gene from the father, located on the opposite chromosome, both of these may end up on the same chromosome. This important means of genetic reshuffling, known as "crossing over," was first discovered in the 1920s in the fruit fly. After crossing over, a second crucial event must occur. The number of chromosomes in each cell must be halved in order to form the gamete. Since the new organism is produced by a fusion of sperm and egg, the number of chromosomes per cell would otherwise double with each generation. This reduction division, as it is called, is accomplished by a series of cell divisions during which the chromosome pairs are pulled apart and distributed among the gametes. The end result is the production of a cell, either egg or sperm, that has only twenty-three chromosomes; this is known as a haploid cell, one in which the chromosomes do not occur in matching pairs. Since any one of these twenty-three chromosomes may have been initially of paternal or of maternal origin, and since each one has undergone changes as a result of crossing over, and since half of the total genetic information has been discarded from each gamete, it can be seen that profound genetic changes have taken place during the production of sperm and ovum, so that neither bears close resemblance to any parental cell and certainly not to any grandparental one.

The next step in sexual recombination is mating. Sperm and

ovum, each bearing twenty-three chromosomes, unite. Their nuclei join together and a new cell is formed, which is known as the zygote, the fertilized egg. This new cell, which has the diploid number of forty-six chromosomes, then embarks on the process of mitosis, cell splitting, by which all the cells that make up the tissues of the human body are formed. All sexual reproduction involves some version of these two processes, although not necessarily in this order, and often (particularly among the protozoa) very different in detail.

MATING AND MEIOSIS

Chlamydomonas is a small green flagellate which often reproduces asexually by simple cell division, but may also undergo sexual recombination. Unlike most other protozoa and almost all multicelled animals, chlamydomonas is naturally haploid, that is, its chromosomes do not occur in pairs. Haploidy is common among lower plants, however, many of which, it is generally agreed, evolved from small green flagellates like chlamydomonas.

Most often chlamydomonas simply undergoes binary fission, forming two new, mirror-image flagellates within the cellulose cell wall, which first expands to accommodate them and then breaks open to set them free. Alternatively, however, one chlamydomonas may encounter another which, although it looks absolutely identical, is apparently different, as revealed by their instant attraction to one another. In fact, when two mating types, as they are called, are introduced, there is first a great clumping together of all of the cells. Then they sort themselves out in pairs and hastily discard their cellulose overcoats. They stick together until their membranes begin to dissolve at the point of closest contact. Finally, the two protoplasms flow together. The cell produced by this union, the zygote, is of course diploid, containing twice as many chromosomes as the original mating cells. In order to get back to the haploid number, the zygote begins a process of meiosis which results in four new daughter cells, each with the proper chromosome complement.

The question of haploidy and diploidy is another one of those

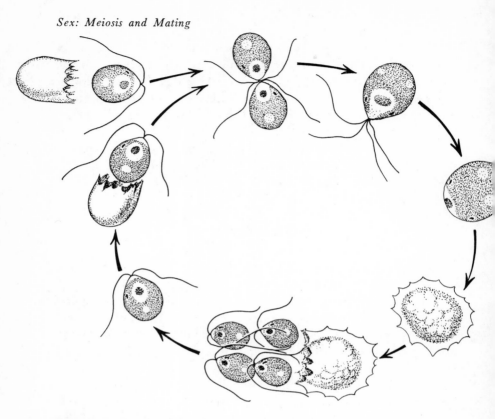

FIGURE 26 Chlamydomonas of different mating strains are drawn to one another at the point of the attachment of the flagella. They fuse, and from this union is produced four new daughter cells. (20 microns.)

issues on which evolution is still undecided. In the haploid cell any serious mistake in the genetic material is immediately eliminated, since that individual cannot survive to reproduce. In the diploid cell, however, there are two sets of instructions, and if one is operative the mistake in the other may be masked for generations and not show up until two similar errors, one on each chromosome of a single pair, are brought together, as in some of the tragic hereditary diseases that turn up from time to time, and often so unexpectedly, in the human species. But diploidy provides for a greater number of genes a greater variety in the gene pool, even if some of these are undesirable. Thus, like sex, diploidy seems destined to endure.

AUTOGAMY

Autogamy means, literally, mating with oneself, and, as the term implies, it involves a reshuffling of the genetic material without anything new actually being introduced. In general, it tends to produce a stable population, like inbreeding among animals. It is practiced by a variety of protozoa—actinosphaerium, for instance. Actinosphaerium is one of a group of rhizopods known as the heliozoans or "sun animals," because their spherical bodies are surrounded by fine radiating pseudopods. Like many of the large rhizopods—and actinosphaerium may grow as large as a millimeter in diameter—it contains a great many nuclei, and it usually divides by plasmotomy, in which the cell simply breaks apart into several pieces. But if actinosphaerium is overcrowded, starved, or cold, it forms a cyst.

As with most protozoa, preparations for encystment are quite

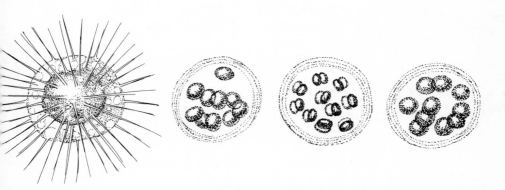

FIGURE 27   Encystment in the "sun animalcule" actinosphaerium. The radiating axopods are withdrawn and the cell retreats into a mucous envelope in which it forms gametes which then fuse with one another. The encysted cell, which is resistant to cold, drying or starvation, will eventually break open, releasing ten new sun animals. (Up to one millimeter.)

elaborate. Actinosphaerium drops to the bottom, withdraws its radiating pseudopods, and its cytoplasm becomes cloudy in appearance as it loses its internal structure. Most of its many nuclei are absorbed into the cytoplasm. It secretes a mucous envelope about itself, and within this envelope it divides into as many pieces as there are remaining nuclei, usually about ten; these look like small black dots set in a yellow jelly. Then each of these "primary cysts," as they are called, divides again, in a reduction division, becoming gametes, like egg and sperm, except that they are all identical. Then, still within these cysts, the gametes fuse, forming once more into diploid cells, the zygotes. The whole process takes about nine hours. Anytime thereafter, depending on warmth, dampness, and food supply, the cyst may break open, releasing ten or so brand-new such animals, each with a single nucleus. The nuclei then divide until a full quota is restored, and the cell continues to reproduce by plasmotomy until ill fortune drives it once more into the safety of a cyst and autogamy.

Autogamy also occurs among ciliates. In these cases, the micronucleus divides within the single animal and the daughter micronuclei fuse and divide again; one of these new micronuclei will develop into a new macronucleus, the old one being discarded. As studies of the demise of tokophrya indicate, a new macronucleus may be extremely desirable. Perhaps among the ciliates the primary function of autogamy is renewal of the chromosomal material rather than diversification.

ALTERNATION OF GENERATIONS

Among the most curious and interesting of the many sexual processes explored by the protozoa are the cycles of the foraminifera. The forams, as they are familiarly known, are sea-dwelling rhizopods that live in shells which they build one chamber, or loculus, at a time. These chambers are often built in spirals so that they resemble snail shells, although there are many species and therefore a wide variety of shells. Chlamydomonas and most other free-living protozoa usually reproduce asexually, with only an occasional sexual encounter. Among the forams, however, species have been found which must reproduce asexually one generation and sexually the next, or perish. In the asexual stage, the internal protoplasm of the foram undergoes schizogony, or multiple fission, to form many small amoebas, each of which on its release immediately begins to make itself a shell. When this individual reaches adulthood, it also divides repeatedly, but this time thousands of small amoebas or, sometimes, flagellates, depending on the species, are formed. These are gametes. These gametes mate with each other or with the gametes from other forams of the same species to produce a zygote. Flagellated forms lose their flagella at this stage and become amoeboid. The zygote then secretes itself a shell. The next generation, the adult form, divides asexually and amoebas are once again produced. The entire cycle takes from about one month to two years in some species, a long time in the life of a protozoan.

The animals that are produced by the asexual cycle are always haploid with only one set of chromosomes and have only a single nucleus, while those produced by the sexual cycle are diploid with paired chromosomes and have more than one nucleus. This was the first example of such alternation of generations, haploid to diploid and back again, ever to be found among animals, although it has been known that all of the higher plants have a similar life pattern. In man also there is, of course, a fleeting moment of haploidy during the life cycle but only in the sperm or the unfertilized egg.

In some species, the young foram that has been produced asexually makes a shell whose first chamber (or proloculum) is relatively large, megalospheric, while the sexually produced foram makes a small, microspheric, first chamber for its shell. Among fossils collected from the ocean floor, some have megalospheric, some microspheric, prolocula, indicating that foraminifera have been caught in the same rhythmic cycle for millions of years.

MALE AND FEMALE

In chlamydomonas and many other protozoa there is no apparent difference between mating types. In some of the protozoa, however, the gametes are clearly different. Those that are smaller, more fragile and more mobile are conventionally designated male, and the larger, receptive cells, female. Vorticella offers a good example.

Vorticella, which sits like an inverted bell or a flower on a long contractile stalk, may divide transversely, producing a daughter cell almost as large as the parent, or it may give forth a tiny bud. Unlike other, nonsexual buds, however, this little creature, the "microconjugant," has no chance for a life of its own. It can survive for about twenty-four hours, and if it has not found a "female" cell by that time it simply dies. A receptive female vorticella looks exactly like any other adult form of the animal, but apparently possesses some special quality that both attracts the male and permits fertilization by it. The microconjugant male lucky enough to find such a female fuses with her, always in the lower third of the body, and then enters the larger cell completely, leaving only its pellicle outside. Next, the male and female micronuclei undergo meiotic divisions within the cytoplasm of the female cell, so that

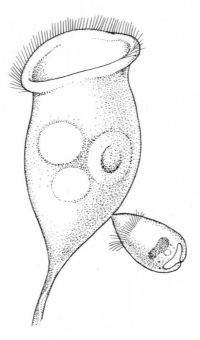

FIGURE 28 Vorticella mating. The smaller cell at the right is the male microconjugant.

when fertilization, the fusion of the nuclei, takes place, the diploid number of chromosomes (four in the case of vorticella) is restored.

Following fertilization, the nucleus of the zygote divides three times, forming eight nuclei, one of which becomes the new micronucleus and seven of which become new macronuclei, which are soon passed on to new daughter cells.

CONJUGATION

Conjugation is an extremely strenuous form of sexual recombination practiced only by ciliates. It has been studied most thoroughly in paramecium, although it seems to be the same in principle, but not in detail, for all ciliates. Each species of paramecium can be sorted into a number of mating types. *Paramecium aurelia* (the one that looks like a slipper), for example, has more than thirty. Some of these can mate with one another, some cannot, and no individuals can mate with other members of the same type. At this point, obviously, all useful analogies between male and female break down completely. When compatible mating types are mixed together, the individuals immediately clump in dense masses and grow in size until, within a few minutes, readily visible groups are formed, which consist of hundreds of adhering ciliates. These sort themselves into pairs, sticking together at first apparently only by their cilia, with attachment likely to occur at any point on their bodies. Since there is no coordination between the members of a pair, they often attempt to swim in opposite directions, one animal dragging the other after it. Sometimes, by chance, they will form chains in which most move in the same direction so that the chains swim like snakes. After about half an hour, individual pairs begin to reorient themselves so they are united symmetrically, first joined by their oral cilia and then by their oral surfaces, which actually fuse during conjugation. They remain in this position for hours, during which a tremendous upheaval of all their nuclear apparatus occurs. The macronucleus shreds apart, disintegrates, and disappears. The micronucleus divides, and then divides again, so there are four in each animal. Three of the micronuclei disappear. The fourth divides again. These events take place synchronously in each of the cells, so that by now each partner has two new nuclei. Then the actual process of genetic exchange takes place. One micronucleus from each crosses the cytoplasmic bridge connecting the two cells. Then each migratory nucleus fuses with the sedentary nucleus, as occurs between egg and sperm in vertebrate fertilization. The two conjugants separate, and each undergoes another round of nuclear divisions with formation of micronuclei and macronuclei. No zygote has been formed, as is the case following most forms of sexual fertilization, and no "child" results. The two individuals that are left are to all outward appearances the same two that existed before conjugation, but actually it can be seen in theory, and has also been proved by experiment, that profound genetic changes have taken place. After conjugation, the ex-conjugants usually undergo several rapid binary fissions during which the new combination of genetic information is reproduced and passed rapidly along.

Conjugation may last as long as forty-eight hours and obviously requires the use of tremendous stores of cellular energy. In forty-eight hours two nonconjugating paramecia could multiply into more than one hundred new paramecia or could consume, alone, some twenty thousand bacteria. Some protozoologists have maintained that without conjugation strains of ciliates tend to age and die out. Others have maintained strains in the laboratory for years and years under conditions where no conjugation is possible, and have found them to be as healthy and vigorous as the conjugating strains. Many have observed the dismal spectacle of some 90 to 95 percent of elderly paramecia of certain strains dying immediately after conjugation. One is tempted to conclude that there must be some reason for conjugation, some adaptive value, since it exacts so high a price in terms of time and energy. But in the absence of proof, one can only say that there is no clear answer at the protozoan level to the classic question of whether or not sex is necessary.

CHAPTER VI

BEHAVIOR

Behavior is a word with an anthropomorphic ring to it, but actually one can talk as reasonably about the behavior of the molecules in a gas as the temperature rises or in water as the temperature falls below 32° F. Some pigments fade, others brighten in the light. Molecules change their configuration in response to warmth. Chemical reactions alter when the acid–base balance—the pH—of the medium changes. It is almost impossible to watch the protozoa for very long without experiencing their actions and reactions in human terms—pleasure, fear, greed, disappointment—but it should be remembered that many patterns of behavior in the one-celled animals, although more complicated, are qualitatively closer to responses of nonliving substances to environmental changes than to our own experience of love or hunger or pain. However, before one dismisses behavior in the protozoa as "just a chemical reaction" or "just a physical response," it is useful to recall that in the last analysis all of our behavior is firmly rooted in this same substratum of physicochemical changes.

In its simplest form, behavior is a response to stimulus, either a positive or a negative response, and in the protozoa this response is generally quite "fixed," that is, it does not seem to be modifiable by experience or "learning." Obviously in some situations a fixed response may not be the most useful one—and indeed, many of our own "learned" responses can be extremely unuseful, as Sigmund Freud pointed out—but in general, behavior, fixed or flexible, is adaptive, helping to preserve the life of the organism and the survival of the species to which it belongs. For example, most enzyme systems are sensitive to temperature; if an animal has a mechanism to guide it to the optimium temperature zone for its particular enzyme systems, it obviously has an advantage over similarly constructed animals which lack this sense of direction. Behavior which is injurious, or "unadaptive," will soon, at least in the protozoan world, lead to the elimination of that particular organism.

BEHAVIOR IN PARAMECIUM

Paramecium is responsive to a variety of stimuli, subtle changes in temperature and chemistry that can be detected by man only by using finely calibrated instruments. But its response to nearly every sort of change is simple and "fixed": avoidance. As paramecium swims forward, it rotates because of the oblique beat of its body cilia, and it does not swim quite in a straight line because the cilia around the mouth beat more strongly than the body cilia. As a result, the rotating paramecium, with its mouth alternately on one side and then the other, swerves like a rowboat in which the oarsman pulls strongly first on one oar and then on the other. Also, these strongly beating cilia create a current which constantly brings to paramecium a sample of the water ahead of it, so it is constantly "exploring" its environment, shifting, and sampling. This can be seen neatly if you put a small drop of particles of carmine dye or India ink on a slide with paramecia. As an animal approaches the drop, a thin stream of particles begins moving toward it, collecting in the oral zone, which is apparently paramecium's testing area. Paramecium's reactions, if it has any, are always negative, and they always result in its turning in the same direction, to the left, which is the side away from the mouth (or aboral side), because of the relaxation of the beat of the body cilia. If the stimulus is weak—such as a slight and unfavorable temperature drop in the water ahead—paramecium, on receiving a sample of the cool water, will turn slightly in its path. If the negative stimulus is powerful, such as a poisonous chemical, paramecium will stop short, reverse its ciliary beat and back up, turn toward the aboral side about 30 degrees, start forward again, testing, and then if necessary repeat this performance. Under a strong negative stimulus,

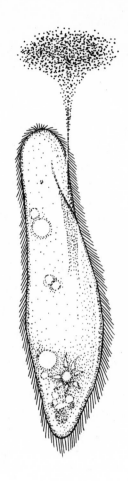

FIGURE 29 The beating of the cilia around the oral groove of the paramecium brings it a sample of the water that lies ahead of it. The dark blot represents particles of India ink which the animal is approaching.

paramecium will turn a full 360 degrees, until a possible avenue of escape is found. Paramecium finds its way around solid objects in the same way. It does not matter which side the stimulus comes from; if the microscopist takes a blunt needle and jabs paramecium on the aboral side, it still turns toward that side to escape, and once it has turned, it will continue in this new direction indefinitely.

The end result of this movement, of course, is about the same as if the animal were "attracted" to a favorable situation. Paramecium, for example, is extremely temperature sensitive; preferring

77

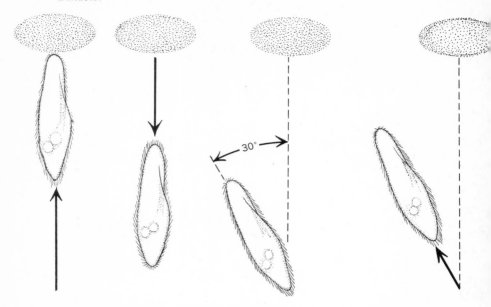

FIGURE 30   If the paramecium meets an obstacle, it backs up, swings 30 degrees to one side, and proceeds forward again. This is a stereotyped reaction which can cause it to rotate a full 360 degrees.

water that is about 80° F., slightly warmer than we like for swimming but colder than a bath. If you place the animals on a slide with water that is below 80°, and then warm one little portion of this slide (you can do this by placing a hot needle on the cover slip), the animals will all gather in this warm spot. They reach it by chance, as they move about through the water. But once they get into it, they literally cannot get out again—no matter how many other warmer and even more desirable spots are available—since as soon as they reach the edge of it, they will receive a sample of cooler water, which will turn them right back again. Similarly, on a slide that contains water that is too hot, paramecia will trap themselves in an area that has been cooled slightly, even only a few degrees.

Paramecia have also been shown to congregate in the same way in an area that is slightly acid, by avoiding alkalinity. This response is also useful, as is the response to temperature. Bacteria, which are the favorite food of paramecium, create a slightly acid environment by their metabolism, and, indeed, paramecium itself creates a slightly acid environment by giving off carbon dioxide, so this response enhances the tendency of paramecia to group together and also to gather where there is food.

Positive responses can also be seen in paramecium. In a bacteria-rich culture, the beat of its body cilia slows down, and when it brushes against an object, particularly one that is fibrous, it tends to cling to it, perhaps with its cilia and perhaps also, some investigators believe, by a slight discharge of mucus. In this way, paramecium's responses permit it to feed and also fasten itself to a bit of decaying matter of the sort around which bacteria are most likely to cluster. Investigators have also shown, by using a bit of filter paper and a weak acid, that the clinging responses occur more readily in a slightly acidic solution. Adding all these factors together, it is possible to see how, on the simplest level, adaptive behavior can evolve in much the same way as other useful physiological reactions.

BEHAVIOR IN AMOEBA

The rhizopods also show avoidance behavior, although they are so differently constructed than paramecium that their reactions are expressed quite differently. If you shine a pinpoint of bright light in front of an advancing *Amoeba proteus*, it will stop, hesitate, withdraw its pseudopod, and form another slightly to one side. If this pseudopod also encounters the light, it too will be withdrawn and another one will tentatively be put forth, until the amoeba finds for itself an escape to an area of darkness. If the light is shone from one side of the animal, the amoeba will steadily move away from the source. If the light is suddenly shone directly on the amoeba, it draws itself together in shock and will even extrude its half-eaten food. If, however, the animal is maintained in light, it will regain its composure and finally go about its normal business of crawling and feeding, just as paramecium will accustom itself to slight changes in temperature or pH that initially elicited its avoidance response. This ability to "make do" is obviously also a behavioral asset.

Amoeba also does not like to be dislodged, and shows a startle reaction to falling, reminiscent of that of the human baby. If it finds itself free-swimming, it will extend pseudopodal arms in all directions seemingly seeking a bit of debris to which to cling so that it can resume its usually creeping, exploratory operations.

The rhizopods, apparently more than paramecium, are capable of highly discriminating positive responses, particularly those involved with food gathering. They can sense prey one hundred microns or more away and also apparently receive some sort of information about the type of victim that awaits them because the size and shape of the pseudopod—in *Amoeba proteus* and many other rhizopods as well—clearly indicate that it has some awareness of the size and activity of its intended meal. Or to put it in another way, a large or actively moving object stimulates an amoeba in a different way than a small or quiet one. A clue to the mechanism of this stimulation can be found in the changes that can be induced in the shape of amoebas by exposing them to different chemicals. This reaction is obviously very precisely localized, as well as very sensitive; the pseudopod extends directly toward the prey while the rest of the animal's body is immobile.

So stimulated, *Amoeba proteus* will remain in patient pursuit of a victim for a relatively long period of time. Euglena, for instance, when it encysts, forms itself into an almost perfect sphere, which will roll at the slightest touch along a smooth surface, and amoeba will follow this elusive sphere until the cyst either rolls completely out of its vicinity or is seized by a more agile predator.

SOME CHEMICAL RESPONSES

Mating cells exert strong attractions toward one another, presumably by some chemical emitted into their watery world. The two tiny green gametes of chlamydomonas find one another in this fashion, just as the male vorticella discovers the receptive female. Recently a similar phenomenon has been discovered in the little suctorian tokophrya. Conjugation, commonly practiced among ciliates, presents a special problem for the suctorians since they spend all their adult conjugable life anchored firmly in the same spot. Recently it has been learned how tokophrya solves this dilemma, and it is presumably by means of some subtle aphrodisiac. When two mating adults are placed close together in the absence of food, their rigid pellicles, which ordinarily look as stiff as glass, begin to waver in shape. Both animals elongate and within several hours begin to develop amoeboid projections. Finally, stretching out pseudopodal arms toward one another, they meet and conjugate. Lest this sound the least bit romantic, it is necessary to add that the entire process can be interrupted instantly by the introduction of an edible ciliate.

A chemical response of quite a different sort is seen in didinium. Didinium is the little barrel-shaped animal that lives exclusively and voraciously on paramecium. The didinia withdraw into a cyst when they have exhausted the paramecium population. Then, mysteriously, they re-emerge as soon as the supply of paramecia is restored. Investigators have shown that didinium can be coaxed out of its cyst by adding a culture of paramecia to the water containing the dormant didinia, presumably as a result of a chemical signal. Such cultures are usually kept in water in which hay has been boiled, since this produces an environment extremely suitable for paramecium and much like its natural habitat. And now observers have found that the hay infusion alone—without a single paramecium in it—will provide the chemical signal that tricks didinium into emerging, somewhat like Pavlov's dogs salivating in response to the ringing of a bell.

BEHAVIOR IN STENTOR

Stentor feeds like a vacuum cleaner, sucking in particles whirled toward it by the vortex created by its beating wreath of heavy, fused cilia. To the casual observer, stentor does not appear to be a fussy eater, accepting, as it does, microscopic bits of inorganic matter along with the nutritive bacteria, algae, and small animals. Sometimes the vacuole containing indigestible debris is short-circuited right out the anal pore. Occasionally, if a particularly large and indigestible particle is caught in the vortex, the whirling stops momentarily, the current is reversed and the offending morsel shot back out again. Laboratory studies have shown, that stentor actually can exercise very fine discrimination in feeding. For example, stentor will reject the majority of sulphur particles, bits of glass or starch grains, even though they are about the same size as the bacteria on which it normally feeds. In one experiment in which stentor was fed chlamydomonas, euglena, and particles of carmine dye in equal mixtures, it ingested, according to the investigator's estimate, fifteen hundred chlamydomonas, eighty-five euglenas, and ten carmine particles, indications of a very clear-cut preference. Selection and rejection take place simultaneously, so rejection is obviously a finely coordinated movement, probably just a flick of a cilium.

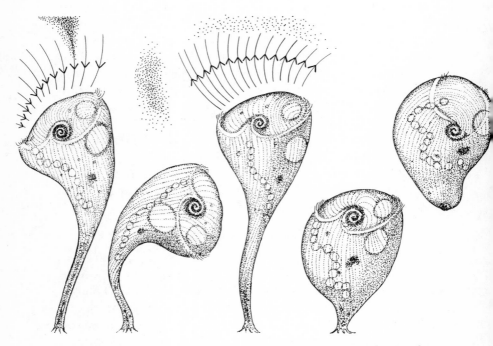

FIGURE 31 Stentor, irritated by a cloud of ink particles, will first bend away. Then, extending again, will reverse its cilia to blow the particles away. If this is not successful, the cell may then contract, loosen its hold, and sail away.

Stentor, when swimming, shows avoidance reactions much like those of paramecium. For example, light evokes a negative response in stentor (although in few other ciliates) and stentor will turn away constantly from light, so that an entire culture of the animals can be captured in one little pinpoint of darkness. Apparently the anterior end of stentor is more light-sensitive than the rest of the body, for stentor will orient itself with its tail pole toward a steady source of light, just as *Amoeba proteus* moves steadily away from it. Stentors that have had their heads removed—a condition which stentor can survive with surprising equanimity—do not respond to light. This is in contrast to *Amoeba proteus,* in which every part of the body appears to be equally sensitive to stimuli. Curiously, one species of the animal, *Stentor polymorphus,* sometimes has symbiotic algae. When it does not have these algae it avoids light, like other species of stentor, but when in possession of its algae it seeks out the light, although the algae have no motile power of their own. In a mixed culture, the colorless stentors will collect in the darkness, those with the algae in the light. Similarly, *Paramecium bursaria,* which is indifferent to light if it does not have its algae, will congregate in it if it does. One can only imagine that the process of photosynthesis creates an internal environment so acceptable to the ciliate that its moving from the light elicits the avoidance response, a stronger response in stentor than the light-avoiding one. So conflict appears to enter into behavior even on the protozoan level.

Stentor attached and feed:ng shows responses that are less stereotyped than paramecium or than free-swimming stentor. If it is poked gently on the side with a needle, it will often bend its huge stentorian head over to suck in the offender for sampling by its cilia. A hard poke, on the other hand, will bring forth the avoidance response, making stentor bend to one side, always toward the aboral side, regardless of where the poke came from. If stentor is poked repeatedly it will contract its body down to the tail pole and then extend again. If the poking continues, stentor will probably contract again, perhaps several times, but then will either ignore the jabbings and continue its primary business of feeding, which it cannot do in the contracted position, or it will detach itself and swim off to some less annoying habitat. Whether or not it leaves appears to depend, at least to some extent, on the richness of the food supply.

Similarly, if stentor is annoyed—by a cloud of ink particles, for instance—it will at first bend, perhaps repeatedly, to the aboral side; if the offensive stimulus still continues, it will reverse its cilia and start blowing the particles away. If bending and blowing are not successful it contracts and waits a minute before it stretches forth again. Once it has contracted, it does not bend or blow again, but may draw back several times and reach out to sample, before it finally gives up, hoists its fastenings, and sails away. It is more likely to give up in a hurry if its site had not previously proved to be a good feeding ground. In other words, stentor's behavior is not entirely fixed and automatic, like the avoidance response in paramecium, but there seems to be some room for "choice," a possibility for spontaneity, for "decisions," and even for mistakes.

BEHAVIOR OF STICHOTRICHA

Because of this question of spontaneity—whether or not a one-celled animal has any choice about what it does—the behavior of stichotricha is especially thought-provoking. This little ciliate has an elongated, flexible body, which is drawn out into a long neck or, more properly, a proboscis, since the cytostome is at its base. When the animal is stretched out, more than a third of it is proboscis. On this necklike nose is a row of heavy, fused cilia and, at its base, a membrane that wafts particles into the extensible mouth. The proboscis curves slightly and the entire body is somewhat spiraled and lopsided.

Stichotricha inhabits the empty cells of leaves of certain common pond weeds. These cells are admirably suited to the ciliate, being just about the right size to fit in snugly and with a convenient small aperture at the top. Stichotricha can retreat into the empty cell completely by contracting its long nose and curling up its body, molding its shape to fit the cell walls. When it is feeding, it braces its rounded posterior against one or more sides of the cell wall by means of its body cilia. Its long extended proboscis projects to one side out of the aperture, and the beating of the cilia set up currents

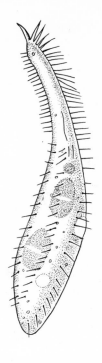

FIGURE 32 Stichotricha, a small ciliate that lives in the empty cells of pond weeds. Its cytosome, or mouth, is actually at the base of its long necklike anterior. (40–170 microns.)

powerful enough to bring particles from some distance to the exposed cytostome. Stichotricha selects its menu from these particles, which may include surprisingly large flagellates, which are swallowed with great effort but apparent determination. The anal pore of the ciliate is in the middle of its back, and the animal rises just far enough out of the cell to allow the fecal material to be deposited outside.

An interesting problem arises when stichotricha divides, which it does transversely, like other ciliates. This means, of course, that the posterior daughter, opisthe (see page 56), finds itself crowded down into the bottom of the plant cell. The anterior daughter, proter, which retains the parental proboscis, membranelles, and cytostome, begins feeding again almost immediately after fission, but the inferior opisthe stays in the bottom of the cell, extending and retracting its new proboscis and apparently feeling the walls of the common habitat.

Then questions of choice seem to arise. Opisthe may remain in

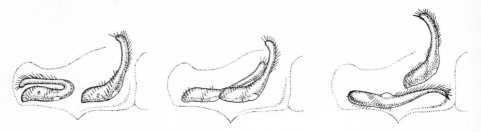

FIGURE 33 After cell division, the pond weed inhabited by stichotricha may become overcrowded. The daughter which migrates is always the bottom cell, the opisthe, which pushes past the superior cell and then finds another habitat. After Froud.

the cell if it seems to be large enough, with both of the ciliate twins extending their heads through the aperture and feeding. Sometimes as many as four ciliates may come to inhabit a single plant cell, their heads emerging like fledglings from a nest. It is more usual, however, for opisthe to migrate after division. Before it leaves, it makes many excursions to the cell opening, pushing its superior sister aside, then retreating, then coming forward again. These movements continue for some time, with opisthe apparently becoming progressively bolder and emerging tentatively farther and farther through the opening. Finally it squeezes its broad posterior through the aperture, past its sister, and escapes.

If there is plenty of pond weed in the immediate vicinity, opisthe will crawl over the plant in a persistent and exploratory manner, as if seeking a new home. It progresses jerkily forward and then, reversing its cilia, backs up, turns slightly, and starts off again. Its long nose is continually poking into crevices and empty cells. Sometimes it goes as far as to enter one, feels its way around inside, backing, twisting, apparently testing, and sometimes sticking its head out the hole as if regarding the neighborhood. Sometimes it is satisfied and stays, but often, particularly if pond weed is abundant, it will enter and leave several cells before settling down. No feeding at all is done until a lodging is secured.

If pond weed is scarce, on the other hand, opisthe begins its life of freedom quite differently, shooting out through the open waters at top speed. Because of the curve of its proboscis, the body swerves as it spirals, so that a large area of water is covered. This period of freedom and very rapid motion may last for several hours. When a suitable patch of pond weed is encountered, the stichotricha begins its quiet, fussy house-hunting, crawling, poking, and testing until it settles down. And even after all this seemingly laborious selection of a rooming place, stichotricha will leave its cell at least once a day to seek a new one, following this same pattern.

It is, of course, impossible to state that this behavior is truly "exploration," that stichotricha is really seeking a nesting place by trial and error, comparing one cell to another and deciding on the most suitable, although it surely appears to be doing so. But it is certain that this particular protozoan, and perhaps others as well, is capable of quite elaborate behavior, certainly unpredictable in its details and far removed from the dance of the molecules heated in a gas or the fixed avoidance movements of paramecium.

BEHAVIOR OF EUGLENA

If you leave a culture of euglena near a sunny window, a clearly visible green cloud will form in the water, and this cloud will move as the light changes, always finding itself a spot which is bright but not too bright. Euglena is able to move toward the light because of a set of special organelles that function as an eye. The eyespot, or stigma, of euglena is clearly visible, as it is in most of the chlorophyll-bearing flagellates. In euglena, it can be seen as a small speck of orange; this orange pigment is a form of carotene, the same sort of substance that colors the cells of a carrot and the yolk of an egg, and that constitutes a fundamental ingredient of the visual pigments of the human eye. It is located on the dorsal surface, or back, of euglena, just over the reservoir, the gullet-like opening in the anterior end of the cell. Arising from the base of the reservoir are two flagella, one long and free, and the other rudimentary. The short segment of the free flagellum that directly underlies the eyespot is slightly thickened. This is apparently the photoreceptor. When the light is shining from one side, as at a sunny window, this photoreceptor is alternately shielded and ex-

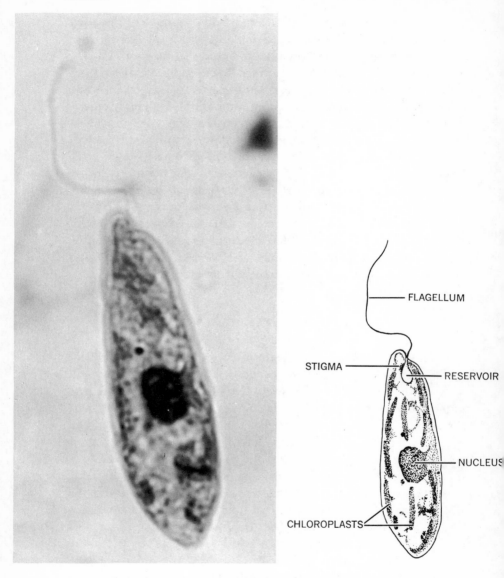

FLAGELLUM

STIGMA

RESERVOIR

NUCLEUS

CHLOROPLASTS

FIGURE 34 *Euglena gracilis,* a common flagellate that often forms a green scum on the waters of ponds or slow moving streams. Its bright color comes from the chlorophyll packed in its numerous chloroplasts. The eyespot, or stigma, is bright orange. Photograph courtesy Ward's Natural Science Establishment, Inc.

posed to the rays by the stigma as euglena rotates, unless the animal is moving directly toward or directly away from the light. In these two situations, the photoreceptor is continuously exposed. Euglena will orient itself into such a position just as an airplane orients itself on a radio beam. Apparently the light acting on the photoreceptor directs the angle of beat of the flagellum, and so determines the motion of the entire animal.

The mechanics of euglena's orientation was discovered long before the nine-plus-two structure of the cilium and the flagellum (page 25) was revealed by the electron microscope. Since that time, this clear structural pattern has been found in many sensory organs, in both the vertebrate and invertebrate world. These include the delicate chemoreceptors of insects, the touch receptors on the suckers of octopuses, the photoreceptors of the starfish (the little red spots at the end of each point of the star) and of the bright blue eyes of the scallop, the mechanoreceptors of the vertebrate ear, the exquisitely sensitive olfactory receptors of the rabbit, as well as those of man. This same structure has been found in the human eye and many other eyes as well. This is especially provocative to those who like to speculate on evolution, for many believe that the eye was the forerunner of the brain—that in the evolutionary course of events the brain did not send forth an eye to see with, but the eye, in itself a primitive brain, called forth the development of a complex brain for the interpretation of what was seen.

In any case, the rods and cones, the light-receiving cells of the vertebrate eye, are divided into two different structural and functional parts. One end of the cell, the one that contains the visual pigments, is the light receptor, and the other end transfers these light impulses into the network of cells, which both resemble and lead into the nerve cells of the cortex. And these two different parts of the cell are connected only by a bundle of fine cilia. In other words, just as the nine-plus-two cilium somehow directs the impulse along the free flagellum of euglena, the same structure transmits the impulses received by our own, so apparently different, eye. Thus may be seen in the bright spot of orange and the trailing flagellum of euglena the starting materials for another long evolutionary experiment of which we have been very much a part.

CHAPTER VII

THE FLAGELLATES

In a single drop of water from the surface of a pond or the sun-touched corner of an aquarium, one can catch a glimpse of what the world might have been like perhaps a billion and a half years ago. In this microdrop there will probably be three or four patches of algae, delicate green or brown latticeworks, and, darting through the water so swiftly you can hardly make out their forms, some tiny swimming specks. If your microscope is good and your eyes are sharp, you can just barely see one or two fine short fibrils beating the water in front of them. These animals—some prefer to call them plants—are small flagellates, the "whip-bearers." No one knows of course—no one was around for another 1,495,000,000 years or so—but some of these modern creatures are thought to resemble closely the beginnings of animal forms on earth.

The earth itself is estimated to be only four and a half billion years old, a fleeting moment in the vast history of the universe. When it first condensed from streaming particles of dust, it fell into an orbit around the sun that was neither so close that organic molecules would be disrupted by the heat, nor so cold that chemical processes would be frozen into immobility. As the earth cooled, seas developed over its surface; indeed, much of the surface that is now land was once covered by water. And over these seas, because the earth was large enough to create a sufficient gravitational pull, an atmosphere formed. All of these factors were essential in setting the stage for the creation of life.

The atmosphere of the young earth was made up of a number

of different gases, methane, ammonia, hydrogen, carbon gases, water vapor, but no free oxygen, such as there is in our atmosphere today. The energy of the sun's rays, of electrical storms, and of volcanic heat acted upon these gases, forming simple organic compounds, molecules containing hydrogen and carbon. As these molecules collected in the water, more complex aggregates were formed. Some combinations were more stable than others; these tended to grow and probably to segregate off into little separate droplets. The droplets may have developed a film around them foretelling the cell membrane. Then—and it must have been sheer chance— one of these combinations began to replicate, to make more of itself and to produce more, identical droplets. It is not known how many times this remarkable event took place. It is known, however, that only one combination of basic chemicals endured. The more that is learned about the biochemistry of cells, the clearer become the closeness of the relationships binding all of them.

These earliest cells probably resemble that group of present-day bacteria, the anaerobes, that live without free oxygen. Bacteria, including the anaerobes, are simpler than most cells; their genetic material is not organized in the form of chromosomes and surrounded by a nuclear membrane—rather it is a long strand of DNA which floats free in the body of the cell. Also, although many of their enzymatic processes are the same as those in the protozoa or in the cells of our own bodies, the enzymatic machinery is not organized in complicated organelles such as mitochondria and chloroplasts.

It is probable that the earliest bacteria-like cells lived saprozoically, nourishing themselves on the soupy medium in which they arose. Eventually, however—again the product of chance—primitive cells developed the ability to collect and store energy from the sun's rays. This was the second big step, because as these photosynthetic organisms collected carbon from the atmosphere and liberated free oxygen, oxygen for respiration became available. The use of oxygen to convert sugar or starch into stored energy—this is the process that takes place in the mitochondria—is far more efficient than any method available to the anaerobes, and without oxygen and respiration more active and more complicated forms of life probably would never have become possible. Ironically, these early photosynthesizers, our necessary forefathers, were most surely destroyed by the changes in the atmosphere that they helped to bring about.

As reflected in modern forms of life, the next step after the bacteria would appear to be the blue-green algae. These resemble the photosynthetic bacteria in their lack of nuclear organization and of mitochondria, but they are larger and the arrangement of their chlorophyll is more like that of "higher" plants.

THE VOLVOCIDAE

The first true cells to evolve were probably much like chlamydomonas, a common small green flagellate, the tiny speck of life from the drop of sunlit water. Chlamydomonas is pear-shaped, not unlike a droplet, but is highly complex and organized as can be seen in the electron micrograph in Figure 35. Chlamydomonas has a rigid cellulose wall, like the stalk of a flower, as do the other volvocidae, the group to which chlamydomonas belongs. The volvocidae are the most plantlike of the flagellates. The bulk of chlamydomonas' volume is taken up with one large chloroplast which, although irregular in outline, is surrounded by a single membrane. Within this chloroplast is a pyrenoid body, a disclike protein structure that is frequently found in association with chloroplasts among the algae and phytoflagellates and is believed to aid in photosynthesis. Several mitochondria can also be seen in this cross section of the cell.

These little creatures are clearly plants, and it is generally agreed that the many-celled plants evolved from phytoflagellates such as these, which settled down, lost their flagella, and took up a social form of existence. Botanists classify chlamydomonas as an alga, but these flagellates are so clearly related to animal life as well that the protozoologists have firmly declined to yield them over.

93

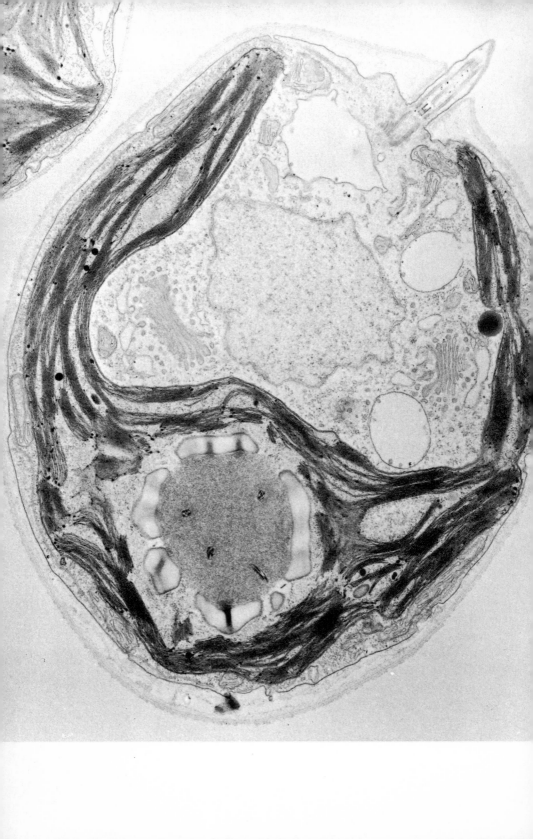

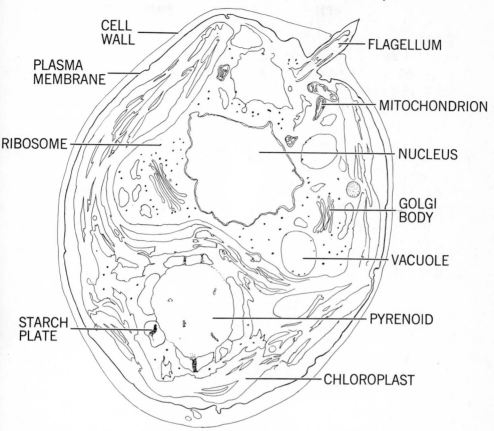

CELL WALL — FLAGELLUM

PLASMA MEMBRANE

MITOCHONDRION

RIBOSOME

NUCLEUS

GOLGI BODY

VACUOLE

STARCH PLATE

PYRENOID

CHLOROPLAST

FIGURE 35   Anatomy of a cell. The photograph (left) is an electron micrograph of the small green flagellate chlamydomonas. (Magnification 20,000 ×.) Photograph courtesy Dr. George E. Palade.

THE CHRYSOMONADS

The chrysomonads, "golden units," are also small chlorophyll-containing flagellates. The cells contain golden-brown pigment bodies, which mask the green of the chlorophyll. Even though each individual is very small, they may multiply so rapidly that they can easily cover the surface of a small pond with a brownish scum. Chrysomonads lack the stiff cell walls of the volvocidae and indeed may often assume amoeboid shapes, putting forth pseudopods, which they may use to move or feed, for many of them are phagotrophic as well as photosynthesizers. It is not unusual for them to lose their flagella and "round up" into nondescript-looking spheres; in this so-called "palmella" stage, they are indistinguishable from algae. Or, some species may lose both their flagella and their chloroplasts and become completely amoeboid and clearly animal-like.

Typically, they have either one or two flagella, and among one of the major groups of chrysomonads the flagella are differentiated, one being always smooth and the other always tinseled with mastigonemes.

The chrysomonads form an important part of the microplankton of salt water as well as of fresh water; they are often found together in large, clearly visible colonies, moving as a mass on the ocean. Among the marine forms, some construct plates of calcium carbonate, known as cocoliths, and some have beautifully latticed skeletons of silica. These shelled and skeletoned forms occur in many different shapes, some very beautiful, ranging from flat discs to stars to plates ornamented with long trumpet-like extensions.

THE DINOFLAGELLATES

Plankton is made up of a great variety of organisms ranging from tiny algal cells so small they slip through the finest plankton nets to shrimp and other multicelled animals clearly visible to the naked eye. It is, however, the chlorophyll-bearing photosynthesizing cells that form the "great meadow of the sea," and that directly or indirectly provide the foodstuff for all the ocean animals.

Among the chief contributors to this phytoplankton, as it is called, are the dinoflagellates. Some of the dinoflagellates are naked but others have cellulose walls, which may give them exotic shapes, such as the widely abundant *Ceratium tripos,* which looks like the anchor for some infinitesimal imaginary craft, or may make them

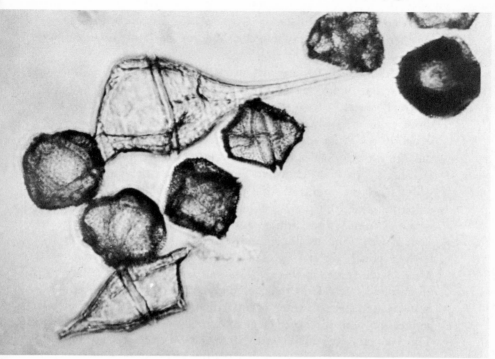

FIGURE 36   Three species of dinoflagellates. The symmetrical organisms are specimens of the luminescent organism *Gonyaulax polyedra,* while the horned objects are two different types of ceratium. Photograph courtesy University of California, San Diego, Scripps Institution of Oceanography.

appear as if they are wearing coats of armor, like gonyaulax. They are sometimes green but more often red or brown in color because of other pigments that mask, as in the chrysomonads, the green chlorophyll. When they form in large numbers they color the sea red for miles around; indeed, it is because of the red dinoflagellates that the Red Sea gained its name. Red tides often form along the coasts of Florida and southern California because of the bloom of the dinoflagellates, and oceanographers use these moving meadows of red to trace the course of great ocean currents. Like the chryso-

97

monads, the dinoflagellates may on occasion become algal-like, losing their flagella, or may lose their chloroplasts and become amoeboid.

The dinoflagellates are also often responsible for the "burning
of the sea at night," as this mysterious phenomenon was called by
the early ocean voyagers. Anyone who has lived or traveled by
the sea can probably recall nights when the touch of an oar or the
splash of a pebble sparked an incandescent glow in the water, or
the churning propeller of an ocean liner seemed to set the entire
sea alight. The first discovered luminescent protozoa was noctiluca,
which is a large (pinhead-sized), atypical dinoflagellate, shaped like
a round apple, the stem of which is a short flagellum. Noctiluca,
which is both photosynthetic and omnivorous, floats in the water
stem down, using its finger-like flagellum to waft particles into its
feeding pouch, in the base of which the flagellum is anchored.
Most of the center of this large, round cell is a vacuole filled with
an extremely acid fluid that is lighter than sea water. The nucleus
is suspended in the center of this vacuole by cytoplasmic threads
connecting with the outer layer of cytoplasm near the cell surface.
This vacuole gives noctiluca the buoyancy it needs to bob near the
surface of the water. Recently scientists have found that the cell,
because of its acid interior, is like a small storage battery. Electrical
discharges pass from the outer cell membrane to the membrane
surrounding the acid-filled vacuole and in so doing set off sparks
of light within the cytoplasm; these sparks discharge rapidly, one
after another, like firecrackers on a string. There are as many as
ten thousand separate light sources within one single cell.

Gonyaulax, a typical armored dinoflagellate, is (like noctiluca)
both red and bioluminescent. Tiny crystals (visible only under the
electron microscope), which are believed to be the source of the
light, have been isolated from one species, *Gonyaulax polyedra.*
Scientists have shown that these isolated crystals, aptly labeled
scintillons, can be made to emit bright flashes of light of the same
intensity and duration as that seen in the living animal.

Another species, *Gonyaulax catanella,* produces an extraordinarily powerful toxin—currently under study by the chemical war-
fare division of the Army. It is for this reason that "red tides,"
which indicate the presence of dinoflagellates in great abundance,

are associated in some localities with beaches strewn with dead fish. Mussels feed on *Gonyaulax catanella,* along with other microplankton, and concentrate the toxin in their internal organs. Although it is not poisonous to the mussels, it is to vertebrates that eat them, including man. For example, one gram of the poison, which is an extremely potent nerve toxin, could kill five million mice in fifteen minutes. For centuries the Indians of the Pacific Northwest posted sentinels to watch for the flashing of the sea lights as a signal that the mussels, a staple of their diet, had become mysteriously poisonous and to warn strangers of the danger.

Many other animals, particularly ocean-dwellers, are bioluminescent. Certain deep-dwelling fish, for example, possess luminescent spots on their heads that not only resemble but appear to function as headlights. Female fireworms swarm in the evening waters after the full of the moon to lure the males by their pooled luminous glow. The familiar flashes of the firefly are also mating signals, precisely timed to attract the appropriate sex and species. The lights of noctiluca and gonyaulax do not seem to have this sort of function, however, nor does the bioluminescence frequently seen in bacteria. Biologists now believe that bioluminescence may have had its origins, in the course of evolution, as a method of eliminating surplus energy. Probably many more organisms were bioluminescent millions of years ago. Some, like the firefly, found a secondary use for it; others discarded it in favor of more efficient energy-storing systems; but in some few, such as noctiluca and gonyaulax, it may remain as an echo of these earlier times.

Gonyaulax polyedra, biologists have found recently, is not only a light but a clock. Like many plants, it reaches its high point of photosynthesis about the middle of the day and, like many other protozoa, including euglena and paramecium, it is most likely to divide in the hours before dawn. Its peak of bioluminescence occurs in just about the middle of the night. These biological rhythms in themselves are not unusual, since many living things, including man, have rhythms involving rest and various kinds of activity; however, in gonyaulax the rhythms are not triggered by environmental factors but are entirely under the control of the organism itself. If gonyaulax is kept in the dark, photosynthesis stops, but the cells continue to show a peak of scintillation at about midnight.

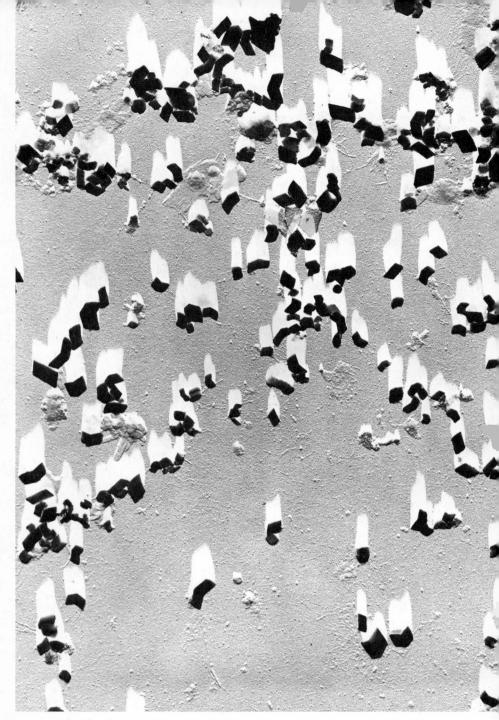

FIGURE 37 Flashing particles, named scintillons, which are involved in bioluminescence, were isolated from *Gonyaulax polyedra*. (Magnification 18,400 ×.) Photograph courtesy *Science* (v. 141, September 27, 1963) and Drs. Richard DeSa, J. W. Hastings, and A. E. Vetter.

If they are kept in continuous dim light, all three cyclical activities continue for many years and many generations. Exposure to light— the creation of an artificial day—can cause gonyaulax to reset its clocks, but once reset they continue to show these new daily rhythms indefinitely when returned to their continuous dim existence. The name scientists use to describe these rhythms is circadian ("about a day"), since most of them follow an approximate rather than an exact twenty-four-hour schedule. Biological clock watchers consider the circadian rhythms to be of great importance, since they indicate that they are governed by the animal's own time-measuring system rather than dependent on some unidentified environmental factor, which would be more likely to have exact twenty-four-hour periods. There are plans afoot to send *Gonyaulax polyedra* up in a satellite to see if it can still keep its time sense. One suspects it can. Microscopic fossils indicate that gonyaulax has persisted virtually unchanged since the Jurassic period, some 170 million years ago, long before man himself existed, much less a man-made timepiece or spacecraft.

EUGLENA

The euglenoids include species that are clearly and permanently plants—that is, photosynthetic—and also forms that feed by phago-cytosis, and so are clearly animal-like—yet the unity of the order is undeniable. Typically, the euglenoid has an elongated body at the anterior end of which is a reservoir, or pocket, into which the contractile vacuole opens. A single flagellum protrudes from this reservoir, anchored at its base, and in those forms that have eye-spots (which includes all the photosynthetic forms), the eyespot is located at the base of this flagellum. Most significant of all, from the taxonomic point of view, is that all the euglenoids store their food reserves not in the form of starch, like plants, or in the form of sugar, like animals, but in a special starchlike material known as paramylum, which is found uniquely among the protozoa of this order.

Euglena gracilis is the species of euglenoid most familiar to the amateur and also one of those most thoroughly studied by the professional. Its anterior end, or prow, which contains the reservoir, is transparent and colorless, except for the orange stigma, the eyespot, but the rest of its body is filled with bright chloroplasts that catch the light and glitter like a kaleidoscope of emeralds as the cell rotates. Euglena is a common contributor to pond scum and the bane of swimming pool owners. Some species are red, and some change from red to green with the changing light, so that the water may bloom red at sunset and green the following dawn.

A common family trait among the euglenoids is a curious motion called *metaboly,* which is an unfortunate name since it sounds like metabolism but has nothing to do with it. From time to time the cell, which appears quite rigid, stops its gyrating forward motion and gives a curious wriggle, a rippling peristaltic motion that begins at the anterior end and travels down the body. Euglena has what appears to be a firm pellicle, and it was difficult for earlier workers to explain how it is able to writhe in this fashion. Electron micrographs show, however, that the pellicle is ridged. These ridges spiral down the body, with soft areas in between; these ridges, the microscopists speculate, may well slide over one another, like the sections of shell in the tail of a lobster or in the outer skeleton of an insect. Some forms of mud-dwelling euglena have lost their flagella and move about by wriggling, but the usefulness, if any, of metaboly to the water-dwellers is not known.

The cells of euglena can be grown in the dark in a medium that contains an organic source of the carbon they usually collect from the air for photosynthesis. These cells are white or bleached in appearance; they contain the basic organelles in which the chlorophyll is arranged, but these plastids, as they are called, do not turn green and become true chloroplasts until the culture is exposed to light. Colonies of euglena that have been maintained in the dark for several years still retain this ability to turn green when exposed to light even though the cells, obviously, are separated by many generations from their light-dwelling, photosynthesizing forebears.

In some strains of euglena, the chloroplast system is very unstable, and even under normal culture conditions—in the light—about

1 or 2 percent of the cells spontaneously lose their capacity to "green." Exposure to heat, to ultraviolet light, or to the antibiotic streptomycin can induce similar changes, or mutations, in almost all types of green euglena. These permanently bleached cells often die out but they may persist under favorable conditions, and their persistence leads one to imagine that some sort of similar mutation might have turned a tiny flagellated "plant" into a protozoan.

The question of how organelles like chloroplasts or mitochondria reproduce themselves has long been a matter of interest to cytologists. At one time the chromosomes of the nucleus were conceived of as directing the formation of structures in the cytoplasm. Very recently, however, using the unstable chloroplast system of euglena —and capitalizing on the fact that euglena, unlike most other plants, can live whether it has functioning chlorophyll or not— scientists have been able to show that formation of the chloroplast is not dependent on instructions from the nuclear control center, but that rather the organelle contains its own control center, in fact, its own DNA. This DNA directs replication of the chloroplasts; and mutations in this DNA, rather than in nuclear DNA, are responsible for the development of the bleached strains. A similar self-replicating DNA system has been found in mitochondria. These bodies were long suspected to be self-duplicating on the basis of both light microscopy and electron micrographs. The way in which a mitochondrion makes another mitochondrion is still unknown, however.

In cells such as euglena, the mitochondria and chloroplasts, of which there are many, do not necessarily replicate at the same time as the nucleus. One tiny phytoflagellate, micromonas, has been studied recently. It contains only one nucleus, one chloroplast, and one mitochondrion crowded into its small pear-shaped interior. Electron micrographs of this organism at various stages of division clearly show nucleus, mitochondrion, and chloroplast—all dividing independently and in perfect synchrony.

OTHER EUGLENOIDS

Euglena has many close relatives. One of the most spectacular of these is phacus. Like euglena, phacus possesses a single motile flagellum that causes it to gyrate through the water, a pellicle with slightly spiraling striations, and bright chlorophyll-containing bodies. Unlike euglena, however, it is broad and almost flat, shaped like an elm leaf or a valentine heart with its edges curling under just enough to make it slightly convex. Phacus spins through the water like a bright, whirling green disc.

Another euglenoid is the colorless peranema. Peranema resembles euglena in its general shape, its single flagellum, and its curious wriggle. In peranema this wriggle is usually associated with its flagellum, which apparently serves a tactile as well as a locomotor function, making contact with a possible victim. The reservoir from which the flagellum arises in euglena is sometimes called the gullet —erroneously since it is not used for eating. The corresponding structure in peranema, however, definitely is a gullet through which pass all sorts of victims, often including its cousin euglena, whose slow, lumbering motions make it an easy target for predators.

THE ZOOFLAGELLATES

The zooflagellates closely resemble the phytoflagellates; they have flagella, they divide longitudinally, and indeed the free-living forms often look so much like their photosynthesizing relatives that the amateur can only distinguish them by looking for the telltale speck of green within the little cell. They are clearly separable from the phytoflagellates, however, by one salient feature: they do not store their food in the form of starch or paramylum but as sugar (either glycogen or a glycogen-like substance) as animals do.

Most of the zooflagellates are very small. A common member of the group is bodo, which, though hardly larger than many common bacteria, can be distinguished by its typical quick, jerky flagellate movement. Like the phytoflagellates, many of these small zooflagellates may form pseudopods; some do so only during certain stages of their life cycle, but others are capable of being amoeboid at one moment, walking around on the pseudopods, and at the next moment resuming their droplet shape and paddling off. This sort of behavior dismays the taxonomists. Some have proposed recently that protozoology should abandon any attempt to make a clear distinction between the flagellates and the rhizopods and group them together in a common subphylum, the rhizoflagellata.

Some of the most engaging of the zooflagellates are those that live in loricas. Loricas, which look like delicate transparent vases or wine glasses, are actually of the animal's own manufacture. The

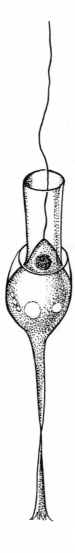

FIGURE 38   A choanoflagellate, or collared flagellate. Food particles, attracted by the beating flagellum, adhere to the outside of the high transparent collar and slide down to its base where they are ingested.

flagellate sits snugly within it with one or two flagella, depending on the species, protruding from the top. Loricas are found also among the phytoflagellates and some of the ciliates. Loricated protozoa tend to live in colonies with the cells arranged on stalks like the buds of fragile flowers or delicate candelabra.

Unique among the zooflagellates are a family of animals with high transparent collars, which are actually extensions of their cytoplasm. Within this high collar a single flagellum beats; this motion sets up a current which attracts food particles. These particles adhere to the sticky outside of the collar and slide down its outer surface to the base where they are engulfed into the cell body. These collared flagellates—choanoflagellates—are of interest to evolutionists because cells of the same type are also found inside the sponge, where they line the feeding chambers, attracting food particles and incorporating them by exactly the same process as the protozoa. Each of these cells of the sponge feeds individually, although it eventually shares its sustenance with other, nonfeeding sponge cells. As a result, even a giant sponge—and some stand taller than a man—is dependent on food particles so tiny that they can be engulfed by the choanocytes. This is obviously an inefficient way to feed a large animal, and the sponge, as it turns out, is generally considered a dead end in evolution.

The majority of species of the zooflagellates are parasitic. Existence as a parasite imposes many unusual requirements, and this group has developed along quite different lines from those of the free-living forms from which they are presumed to have evolved. For this reason they will be described separately, in Chapter XI, along with the other parasitic protozoa, which they more closely resemble in their life cycles.

CHAPTER VIII

THE RHIZOPODS

Amoeba and the amoeba-like protozoa first attracted the attention of biologists because of their seeming primitiveness. The subphylum or class to which they belong was called then—and still is by some protozoologists—the sarcodines. The concept of the sarcode, or "living jelly," has long since disappeared. Yet, in all truth, an amoeba does look at first glance like a drop of primordial ooze.

The most conspicuous common characteristic of the rhizopods is the absence of any rigid cortex or cell wall. In fact, at one time it was believed that amoebas had no "skin" at all but were simply held together by physical forces, like a drop of oil in water. It is now known, however, that the rhizopods, like all other living cells, are surrounded by what is called the unit membrane. In recent years scientists have discovered the physical structure of this unit membrane in terms of the actual molecules that compose it. Under the electron microscope the unit membrane looks like a sandwich, or three-ply sheet, composed of two layers with a less dense filling in between. By a combination of chemistry, electron microscopy, and some brilliant deductions, the "filling" has been shown to consist of two layers of lipid (fat) molecules. Each of these molecules has a water-loving (hydrophilic) and a water-hating (hydrophobic) end to it. If a thin layer of these molecules is spread on a surface of water, they will all line up vertically forming a fine film on the surface, with their hydrophilic heads under water and their hydrophobic tails sticking out.

In living systems, where they cannot get their tails above water,

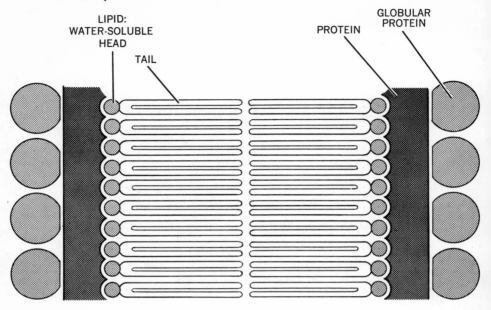

LIPID:
WATER-SOLUBLE
HEAD

TAIL

PROTEIN

GLOBULAR
PROTEIN

FIGURE 39   Diagram of the structure of a cell membrane. The inner layer is made up of lipid molecules and the outer is composed of proteins and polysaccharides.

the lipid molecules tend to form into double rows with the hydrophobic tails of each attracted toward one another and the hydrophilic heads pointed outward. This double row of lipid molecules is the "filling" of the molecular sandwich, the middle layer in the three-ply sheet. Aligned along the hydrophilic borders of the double-lipid layer are the two outer sheets, each of which is made up of large complex molecules, proteins, and complicated sugars (polysaccharides). The unit membrane is used over and over again throughout all cells, enclosing vacuoles and droplets, surrounding various organelles, and making up an intricate network of structures that compartmentalize the cell's activities and provide surfaces on which chemical reactions take place. The exact composition and arrangement of the proteins and polysaccharides vary from cell to cell, membrane to membrane, and even from point to point over the same membrane. These variations in the chemical "texture" of the membranes undoubtedly have a great deal to do with what passes through the different membranes and what takes place on their surfaces.

Coating the outer surface of the outer membrane of *Amoeba proteus* and other larger amoebas is a diffuse fringe of hairlike molecules. These are mucoproteins, a sticky combination of mucus and protein, and are thought to help the amoeba in adhering to the surface along which it travels and in holding fast to molecules to be incorporated by pinocytosis. When a portion of this outer membrane is pinched off during phagocytosis to form a food vacuole, it quickly loses its hairy fringe; also, when *Amoeba proteus* is in the process of division, the fringe seems to become markedly reduced.

One of the provocative questions about the cell membrane is what happens to it in the course of amoeboid movement. Projections of pseudopods are sometimes almost explosive in their rapidity. Does the membrane stretch? Or is it perhaps true, as some investigators believe, that new membrane is forming all the time, in fractions of seconds? Membrane can form very quickly; every time a contractile vacuole pumps, for instance—and this may occur almost as rapidly as a heart beat—the membrane is ruptured and re-formed. Food vacuoles may similarly repair themselves with great haste if they are torn by the movements of the struggling captive.

Amoeba proteus is almost transparent; in fact, its protoplasm is so like that of the surrounding medium and its shape so indeterminate that the beginning microscopist has a hard time finding it the first time. What gives the amoeba away—and what the experienced eye looks for—is the streaming of objects through the cytoplasm. Among these objects are, of course, food vacuoles, some of which may be remarkably large. Amoeba frequently ingests diatoms, large algae that live in glass houses, transparent boxlike containers of silica. Amoeba will swallow a diatom which, container and all, is almost as big as itself; bacteria and flagellates are also favored, and the latter may often be seen thrashing briefly within their vacuoles. The nucleus is large and as it rolls through the cytoplasm it can be seen to be shaped like a disc and slightly concave. There is generally one contractile vacuole, readily identifiable because of its emptiness and its tendency to disappear from time to time as it discharges its contents; the contractile vacuole commonly forms at the rear end of the amoeba, moves forward with the advancing sol, presumably collecting excess water as it goes, and then falls back to the rear again, where it discharges its contents. Amoebas also contain fat globules, recognizable by their oily texture, and a large number of small, darkish crystals. These crystals, which are apparently mineral deposits, are extremely useful to the microscopist; because of their number and regularity, they serve as clear indicators of the flow of sol and gel. But their usefulness to the amoeba is not known.

There are many amoeba-like rhizopods, all of which move and feed by means of ever changing extensions of their cytoplasm and all of which lack any sort of rigid cortex or shell. Some, such as *Amoeba proteus,* project a number of thick pseudopods. Some send out a single "foot" which fans out before them. Others do not form any feet at all, but simply flow themselves forward, like tiny mobile sausages; these are known as the limax amoebas, limax being the Latin word for slug. Many of these sluglike rhizopods, which are usually soil-dwellers, may also develop flagella. In fact, flagella can usually be induced simply by diluting the medium in which they are found some five- or sixfold with pond water. Within a few hours, flagella appear; the new flagellates swim around for a day or so, a kind of remembrance of times past when their ancestors led a more watery existence, and then settle once more on the bottom, where they lose their flagella and resume an amoeboid existence.

These naked amoebas are to be found everywhere—in ponds, in the soil, in the seas, and even, as some few of us have been uncomfortably aware, in the intestinal tract.

THE TESTACEA

We are accustomed to think of the amoebas as shapeless, as of course their name implies—in fact one species is even known, somewhat redundantly, as *Chaos chaos.* Yet the shelled rhizopods, of which there are many, are among the most ordered and exquisite of all cells. The first one of these likely to be encountered by beginning microscopists is arcella, which is commonly found in ponds and on the surface of aquatic plants. Arcella is a member of the testaceans, which are characterized by having a single shell, with no chambers, seams, or joints. Seen under the microscope, arcella appears as a semiopaque disc, almost a perfect circle. In younger animals the disc is a bright lemony-yellow, but it turns brownish with age. When you first see it, the disc does not seem to be alive; in fact, sometimes it is not, for the shell may outlast its occupant, but usually a minute or so of observation will be rewarded by the appearance of small, blunt pseudopods, moving like shadows around the shell rim.

Although arcella appears flat from the top, actually its shell, or test, is a cup-shaped container that surrounds most of the animal. It is concave on the bottom with a round aperture in the middle of the bottom surface through which the pseudopods are put forth. Arcella can raise itself off the ground and stagger around on these pseudopods, but usually it remains stationary and uses them only for gathering food. When arcella divides, one of the daughters retains the old shell and the other secretes itself a bright new one out of keratin-like material, the same sort of tough protein substance which forms our own fingernails as well as the feathers, scales, claws, and hoofs of other animals.

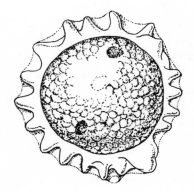

FIGURE 40 Arcella is a small amoeba which is almost covered by a yellow cup-shaped shell. Often it is only slight movements of the small pseudopods extending out from under the shell that reveal the presence of a living animal. (30–100 microns.)

Another common testacean is difflugia, often found in the same environment as arcella. Difflugia does not produce its own shell but rather secretes a sticky organic substance on which then forms a sort of haphazard mosaic made up of particles of sand and, often, the discarded shells of diatoms. Difflugia, which may be very brightly colored, depending on the building materials it finds, is shaped like an upside-down vase or urn, with long pseudopods protruding from the vesicle's circular mouth.

THE HELIOZOA

The heliozoa are a striking collection of animals, with spherical bodies and stiff radiating filaments, so they look like a child's drawing of the sun—hence their name, the sun animals. At first they would not appear to be rhizopods at all, but on closer inspection, one can see the characteristic flow of sol, a transparent streaming, up and down the needle-like extensions. These extensions, called axopods, are still, in contrast with the pseudopods of the amoebas, each being supported by a fine central rod, extending deep into the body of the cell, like needles in a pincushion.

The sun animalcule you are most likely to encounter is actinosphaerium, which floats near the surface of ponds and slow-moving streams, drifting with the currents and occasionally maneuvering by means of its axopods. Actinosphaerium is commonly 200 or 300 microns in diameter and may even reach a millimeter (1000 microns, or 1/25th of an inch), in which case it is visible to the unaided eye as a white speck on the water. The spherical body of actinosphaerium is divided into two distinctly different zones. The inner zone, the endoplasm, is dense and granular and contains numerous nuclei and also food vacuoles, crammed with diatoms and other algae, protozoa, and small crustaceans. The outer zone, the ectoplasm, contains so many vacuoles it appears to be composed entirely of froth. Some of these bubbles are contractile vacuoles; these appear and disappear between the needle-like axopods, swelling up and bursting and swelling again at regular inter-

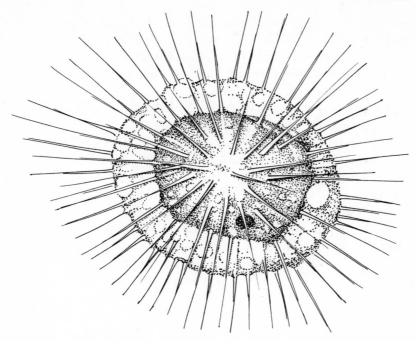

FIGURE 41 Actinosphaerium is a large (almost as big as a pinhead) and handsome member of the amoeba family. It has been called the "sun animal" because of the sharp raylike spines radiating from its spherical body.

vals. The other vacuoles probably help to keep actinosphaerium afloat.

Actinosphaerium usually lies quite still, waiting for its prey. When small moving animals pass near, they stop, as if stunned or paralyzed; apparently the axopods secrete some sort of toxin. Small particles, such as algae and tiny flagellates may be carried down an axopod by the flowing protoplasmic currents. In the case of larger prey, like ciliates or rotifers, the axopod withdraws into the animal, bearing its booty with it. Sometimes several axopods will cooperate in delivering the captive into the body of the cell.

Actinosphaerium usually divides by plasmotomy. Sometimes, particularly when food is in short supply or the weather turns cold, it will encyst, and within the cyst will form gametes, which then pair to form zygotes. The zygotes may remain encysted for many weeks until favorable external conditions somehow coax them to emerge. The young actinosphaerium, freshly excysted, has only a single nucleus, but this soon divides as the cell feeds and grows to its multinucleated adulthood.

Actinophrys sol is another heliozoan, even more common than actinosphaerium, although smaller (about 50 microns) and therefore not so easy to find and watch. Unlike actinosphaerium, actinophrys contains but a single nucleus, quite large and centrally located. The proximal ends of its axopods, the electron microscope shows, are apparently embedded in the nuclear membrane. Actinophrys does not capture food with its axopods but rather sends out a special pseudopod toward its prey, engulfing it and drawing it into the cell. These feeding pseudopods are tailor-made for the intended victim. If the object is small—a speck of alga—a straight narrow pseudopod is briskly extended which, on coming into contact with the alga, forms a neat little cup around it and pulls it in. A large motionless object—a diatom, perhaps—calls forth a broad spreading pseudopodal wave, which expands as it reaches the prey and then washes it back with it into the cell. If the intended victim is both large and active, a big saclike pseudopod is sent out, and once the prey is caught, the axopods close behind it like a palisade of spears to cut off the captive's escape.

Actinophrys is also a floater but, like some of the other heliozoa, can on occasion roll along the bottom like a ball. This is accomplished by a sort of handspring made possible by the progressive shortening of the axopods in the direction of motion, so that the body of the animal is always, in effect, running downhill. A few heliozoa also possess flagella, generally two, as well as their axopods, which provide a less strenuous method of locomotion.

Many of the sun animals secrete a gelatinous capsule around themselves, and in some species this capsule serves as a matrix for the support of sharp fine spicules or flat plates of silica of their own manufacture. Others enclose themselves in elegant, intricately latticed baskets through which their radiating axopods extend and mount themselves on graceful silicacious pedestals as one would mount a work of art.

RADIOLARIA

The radiolaria are best known to us because of their skeletons; on the famous voyage of the HMS *Challenger* in the 1880s the young Ernst Haeckel, who was to be one of the great biologists of his time, collected specimens of almost 4000 species, which he drew in exquisite detail. Another 1000 or more species have been discovered since that time, dragged up in plankton nets or dredged from the ocean floor, for all of the radiolaria are marine animals. They are large, as protozoans go; some of them measure several millimeters in diameter. Some species cluster in colonies clearly visible to the ocean voyager.

The majority of species have skeletons, usually formed of silica, and each different species makes its own genetically determined kind. These are skeletons in the true sense—that is, they occupy the interior of the cell, rather than forming a protective shell around it. The bodies of most of the radiolaria, like those of the heliozoa are spherical, and divided into two zones. In the case of the radiolarians, these two zones are separated by a heavy membrane in which the endoplasm is enclosed. This membrane has anywhere from one to many hundred pores or perforations in it (depending on the species), to provide continuity with the ectoplasmic portion of the cell. Just outside the membrane is a stiff froth of gelatinous material and, coating the entire cell, a fluid layer by which the animal feeds either by axopods, like those of the heliozoa, or filopods, fine unsupported filamentous pseudopods. Mixed in this froth, or sometimes in the endoplasm itself, are tiny yellow-brown specks of small dinoflagellates and chrysomonads in the palmella stage. Some radiolaria float near the ocean surface, but others have been found at depths as great as three miles, far from the photosynthesizing plankton and other small living things that are its natural source of food. Presumably the vacuoles in the frothy ectoplasm enable them, despite the comparatively great weight of their skeletons, either to float at any depth they choose or to rise and sink like submarines.

There are several distinctive types of skeletons. In one, long spines or needles radiate out from the center of the inner capsule extending through the ectoplasm beyond the body of the cell. The points at which these spines leave the body surface are surrounded by contractile fibrils, which can move the spines individually or cause them to withdraw, pulling in the cell membrane with them and contracting the ectoplasm.

A second kind of skeleton, the most common, is constructed in the form of a hollow latticed sphere; in fact, there may often be several such spheres, concentric, balanced one inside the other, like those that Chinese ivory carvers used to make to display their extraordinary patience and virtuosity. Or the skeletons may take the shape of almost any imaginable polygon, a triangular prism (or tetrahedron), a pentagon, a cube, or up through twelve- and twenty-sided figures. Sometimes strange projections—thorns, horns, spears, or tridents—protrude from these spheres and polygons, breaking their geometric regularity.

Skeletons of another type, the most surprising, have only bilateral symmetry. These resemble wicker baskets, carved bells, helmets, the skulls of fanciful tiny monsters, ironwork trivets, or elaborate heraldic devices. As one looks at radiolarian skeletons one is reminded inevitably of the meticulous filigree of snow crystals, but unlike snow crystals—no two of which are ever supposed to be the same, at least according to legend—the skeletons made by any particular species of radiolarian are always the same. Each time a new cell forms, it fashions itself precisely the same skeleton as that of its parents; in fact, as we know from fossil remains, some of the radiolarians have been making the exact same skeleton for millions of years.

THE FORAMINIFERA

The foraminiferans, familiarly known as "forams," constitute another large group of marine rhizopods. All forams bear shells, usually multichambered and of a great variety and are found drifting with the surface plankton, floating at great depths in the sea, or actually dwelling on the ocean floor. The sharp-eyed Leeuwenhoek was the first to record the observation of a foram; he found it in the stomach of a shrimp. You yourself can readily find foram shells among the grains of sand at the tideline, swept in from the ocean bottom where they pile up at great depths. The shells often resemble those of snails and other shelled invertebrates; indeed, forams were first thought to be tiny mollusks. It was Dujardin, the father of the *sarcode,* who first recognized that they were related to the amoebas.

Foram shells vary greatly in their architecture, depending on the species, but they share several common features. They are built one chamber (or loculus) at a time. The neonatal foram lives in one small shell until it begins to outgrow or overflow it, at which time it constructs a second one, and so on, each a little larger than the preceding one. Fossil foram shells have been found, which are several inches long, containing many loculi. These loculi are not separate compartments, as in the chambered nautilus, for instance, which uses its airtight chambers for flotation, but are connected with one another by foramina (which is Latin for "windows" and the source of their name), and the animal as it grows simply distributes itself throughout the many chambers of its mansion. Every time the animal makes an addition to its home, it may add a new layer of shell to each preceding loculus, so that the first loculus, for example, is covered with as many layers as there are chambers in the entire shell. Depending on the species, the chambers may be added linearly, so that the shell simply gets longer and longer, or it may spiral like a snail shell, or successive shells may enclose the previous ones almost completely, like petals on a flower that enclose the inner petals, and leave only the bases of the original loculi in communication with the outside. One particularly common type, globigerina, consists of round chambers stuck together in an untidy spiral to form a lumpy sphere. Foram shells are made of calcium carbonate—lime—which the animals extract from the sea water. The white cliffs of Dover, and similar chalky deposits throughout the world, are in fact the result of the long accumulation of the discarded shells of foraminiferans.

The outer surfaces of all the loculi are covered with perforations through which the cytoplasm extrudes, forming a thin film over the entire outer surface of the shell. From the main aperture, the opening of the last-formed chamber, fine filaments are sent forth. These threads fuse together where they meet, forming a sticky network, which has been likened to a spider's web. Food particles adhere to the web and are carried down into the body of the animal by the constant flow of ectoplasm. Even the most slender of these filaments carries a constant two-way stream of traffic, clearly visible because of the adhering food particles and the intracellular granules. These streams always move at exactly the same speed, carrying the granules along as if they were on a conveyor belt, out to the end of the filament and back again, only occasionally running into traffic jams at the points where filaments cross and fuse.

Only a small percentage of the known species of foraminiferans are planktonic, but these multiply so rapidly that they far outnumber, in terms of individuals, the deep-sea-dwellers eking out an existence in a less favorable climate. The discarded shells of these planktonic foraminiferans have fallen like a slow and steady rain upon the ocean bottom for more than half a billion years. During these years, thousands of species have come and gone. Geologists have long known that identification of these fossil forams has practical applications. In Texas and Oklahoma and other areas rich in oil deposits—all of which were once under the surface of the sea—they are able to identify the stratum reached by the oil drills by studying the cores for fossil shells. More recently, methods have been developed by which hollow metal tubes can be driven into the ocean floor to collect cores more than a hundred feet in length, some of which contain fossil shells deposited millions of years ago. Each of these long cores, it has been found, can be read like a history book divided up into a number of clearly definable chapters, each representing many thousands of years. These chapters represent sudden climatic changes, the result of the great ice ages that caused certain species of forams to disappear entirely and others to take their place. By carbon-dating and other modern techniques, it is possible to tell approximately how long ago these changes occurred. The beginning of the Pleistocene epoch, the time during which the evolution of man and of many other present-day forms of life began, was marked by the first ice age. According to the evidence of the ocean floors, the Pleistocene epoch lasted some one and a half million years, during which three more great glaciations occurred, the last of which retreated eleven thousand years ago, just before recorded history began—only a fleeting moment of time in the long history which is still being written on the ocean floors.

CHAPTER IX

THE CILIATES I
(THE HOLOTRICHS)

Since the time of Leeuwenhoek the ciliates have been the favorites of protozoologists. They are the most appealing to observe, for amateur and professional alike, because they are the most animal-like, even the most "anthropomorphic," with their remarkable varieties of appendages, clearly visible organelles, and seeming seriousness of purpose. Taxonomically they are in far better order than any other protozoa; the ciliata is the only one of the four major groups (actually, there is some question whether or not there *are* four major groups) on which there is general agreement about the various subgroupings. Since taxonomy is, ideally, a reflection of what has happened in the course of evolution, these classifications take on special interest and meaning.

Also, as some biologists have long recognized, the ciliates offer spectacular systems for the study of morphogenesis—the way in which the particular forms and structures of living things come into being. Morphogenesis has been one of the central problems of biology since the time of Aristotle. Now with new instruments and new techniques morphogenesis can be studied at the cellular level. And within a single ciliate the biologists can find the widest variety of cellular structures, some of which are virtually unexplored, many of which take their shape, their *morphé*, within a matter of minutes and before one's eyes. These living structures, the cilia, for instance, are of great intrinsic interest, revealing as they do the way problems of form and function have been solved almost at the molecular level. Their morphology takes on a special and almost awesome meaning, as we find these same structures distributed through an endless variety of living things, marking the relationships that bind us all to one another.

CHARACTERISTICS OF CILIATES

The ciliates have a number of special characteristics that set them apart from the other protozoa. First, of course, they all possess cilia, at least at some stage during their life cycle. These cilia may cover their entire bodies, as in paramecium; or they may be fused and organized into complex organelles for feeding and locomotion, as in euplotes; or both, as in stentor. Second, and concomitantly, they all, and always, possess a system of kineto-somes, the basal bodies which are the origins not only of the cilia but of many other structures as well. Third, they tend to divide transversely, one animal forming on top of the other. Fourth, the ciliates possess two functionally different kinds of nuclei—the micronucleus, which stores the information of heredity, and the macronucleus, which dictates this information to the cell. Fifth, only the ciliates (though not *all* ciliates) practice conjugation, that strange and strenuous process for the reshuffling of their genetic material. So they are indeed a distinct and well-defined group.

The class ciliata is in turn divided into two large subgroups; one, the holotrichs, is the subject of this chapter, and the other, the spirotrichs, is the subject of the next. All spirotrichs possess a complex system of special ciliature which spirals, always clockwise, toward the mouth, or cytostome. The holotrichs are most simply defined by their lack of this complex oral ciliature. Among this latter group, the cilia tend to be uniform and often cover the body of the entire animal. These body cilia typically grow in longitudinal parallel rows, known as kineties. The individual "lashes" are not separate, like the hairs on our head or on our arms, each of which grows in its own separate follicle. Every cilium is part of a network. Extending from each ciliary base,

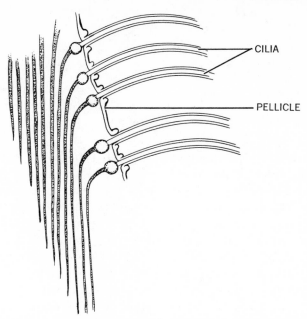

CILIA

PELLICLE

FIGURE 42   The kinetodesmata, showing the cilia, the kinetosomes, and the connecting fibers that lie beneath the pellicle.

the kinetosome, is a relatively short fiber which always leads off to the (animal's) right. These fibers join in a bundle, the kinetodesmata, which runs parallel to the row of kinetosomes. Curiously, none of the fibers runs the length of the kinetodesmata, as they would, for example, in a bundle of telephone wires. Each runs a short distance, roughly between four and five kinetosomes, then stops, so the bundle always contains only about five separate fibers, none of which, according to electron micrographs, actually touches another. There is a temptation to look upon the kinetodesmata as a neuromotor system—surely something coordinates the synchronous beating of the cilia—but physiologists are not sure whether this unorthodox structure could actually function as a useful transmitter of information.

THE GYMNOSTOMES

The various subgroups of holotrichs are based largely on the position and complexity of the feeding apparatus. The simplest of all are the gymnostomes, the "naked mouths." In its most primitive form, the ciliate "mouth" is no more than a thin area in the pellicle in which a food vacuole can readily form. In most modern gymnostomes this simple mouth, which lacks any special external ciliature, is supported on the inside by a circular fence of skeletal rods, the trichites. This fence may be rigid or it may be expandable, as in the case of didinium, for example, which can swallow a paramecium larger than itself. Often the cytostome is located at the anterior end of the cell where the swimming motions of the animal will cause bacteria and other food particles to collect, but other locations have also been tried in the course of evolution. One very common ciliate, for example, is dileptus. The anterior

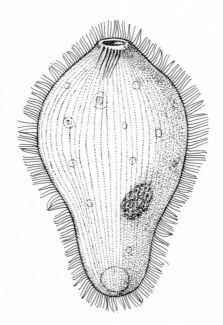

FIGURE 43   Prorodon, one of the gymnostomes, or "naked mouths." These are the simplest of the ciliates. (100 microns.)

portion of dileptus is very narrow and gracefully curved, like the neck of a swan. Actually, this anterior portion is more accurately described as a proboscis, because at its base, just where the body begins to broaden, is found the naked cytostome with its bundle of supporting trichites.

THE TRICHOSTOMES

A slightly more "advanced" division of the holotrichs are the trichostomes (or "hairy mouths"). In these animals, the food vacuole forms not at the surface of the pellicle, as in the gymnostomes, but at the base of a depression in the pellicle, the vestibulum, which is lined with cilia. Colpoda is perhaps the most common ciliate of this sort. Its body is kidney-shaped, and the vestibulum can be seen in the depression on the concave side of the body, just above the midline. The vestibulum leads directly to the cytostome. The body cilia of colpoda emerge from deep furrows in the body and usually occur in quite widely spaced pairs.

Colpoda is unusual among ciliates in that it can divide only when encysted, and in fact spends a good deal of its life moving in and out of cysts, as compared to paramecium, for instance, which has never been found in an encysted form. The reproductive cyst of colpoda is actually only a fine membrane in which the organism divides, usually twice to make four daughters and sometimes three times to make eight. Newly hatched, colpoda has an absolutely clear cytoplasm, but it is such an efficient feeder that within a few minutes it is crammed with as many as two hundred food vacuoles. In addition, it readily forms resistant cysts, usually in response to hunger. These cysts have two walls, one delicate and transparent and the other, outer wall, hard and wrinkled. Colpoda decreases in size before encysting—the cyst is smaller than the free-living ciliate—and then, once inside the cyst, loses its cilia and absorbs its food particles. When conditions improve, the outer cyst ruptures, the animal regains its cilia and rotates briefly within the transparent cyst, which gradually disappears. The newly hatched organism eats and grows, and then, within a matter

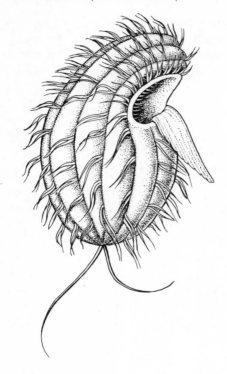

FIGURE 44 Colpoda is one of the most common of the trichostomes, the "hairy mouths." It is very small and, even with the microscope, you would not ordinarily see it in such detail. (30 microns.)

of hours, forms a reproductive cyst in which it once more divides. Sometimes—if the little film of water in which they live should dry up, for instance—colpoda may form a resistant cyst around a reproductive cyst; these resistant cysts may not hatch for years, but when they do there will emerge four to eight fresh, voracious "newborn" gymnostomes.

THE HYMENOSTOMES

The hymenostomes (membraned-mouths) have developed an oral ciliature that is quite distinct from the simple body cilia. One common member of this group is tetrahymena (four membranes), which is a small, widely occurring fresh-water animal. The body of tetrahymena is extremely plastic, so the members of even a pure laboratory stock, all of which should be identical twins, range through a wide variety of shapes from sphere, through egg, pear, and cucumber to sliver. Tetrahymena has a relatively small, roundish opening near the anterior end of its body which leads into the funnel-shaped buccal cavity. On one side of the cavity is an undulating membrane, made up of a long row of cilia that are fused together like the fine fibrils that make up a bird's feather. Opposite the undulating membrane are three smaller membranelles, groups of fused cilia. The beating of these four membranes direct food particles into the buccal cavity and down the funnel to the cytostome.

Paramecium, which is larger than tetrahymena and far easier to observe, is the best studied and most familiar of the hymenostomes, and perhaps of all the ciliate family. On its ventral, or underneath, surface is a long, large oral groove running half the length of the animal's body. When paramecium has room to move freely, it rotates as it swims and this groove is readily visible. The oral groove leads to the broad vestibulum, which narrows into the funnel-shaped buccal cavity leading to the cytostome. The vestibulum is lined with the same type of cilia that covers the rest of the animal's body, while the buccal cavity contains one undulating membrane, curving in an arc along the right wall of the funnel, and three or four bands of cilia. The cilia around the oral groove beat more rapidly than the body cilia when paramecium is browsing. This beating, combined with the forward movement of the animal, serves to channel particles into the vestibulum, and from there into the buccal cavity. The cilia of the vestibulum and the special ciliature of the buccal cavity push the particles along or, sometimes, eject them, although it is difficult to determine exactly what the basis for selection and rejection is. When a sufficient quantity of food—bacteria are preferred—is collected at the cytostome, a vacuole forms. Dangling down from the base of the cytostome is a group of straggly hairs, which seem to form a sort of net, holding the food vacuole as it fills and then releasing it for its trip through the body. When the food vacuole is emptied of its digestible contents, it leaves the cell by means of a special anal pore, the cytopyge, which has a fixed location on the underside of the animal. Food vacuoles are easy to see in well-fed paramecia, particularly if the diet is varied with colored bacteria, green algae, or—apparently equally acceptable though less nourishing—India-ink particles.

Another conspicuous feature in paramecium is the contractile vacuole, of which most species have two. These are shaped like flowers, each with six or seven daisy-like petals leading to a circular center. They pulsate. First the petals fill and swell, emptying into the central vacuole, which swells in turn and then pumps its contents through the cell wall. The large, dark macronucleus is also readily visible, and in some species you can make out the much smaller micronuclei.

The pellicle and subcortical structures of paramecium have been studied extensively using a variety of staining techniques—especially

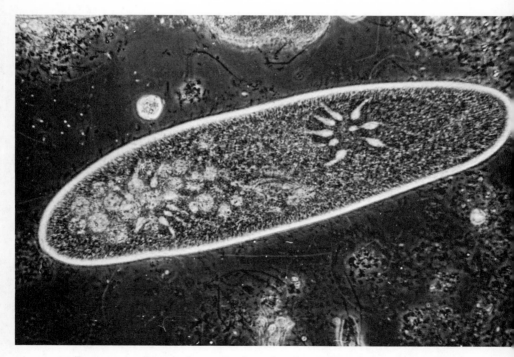

FIGURE 45 Paramecium, showing the flower-shaped contractile vacuole. Photograph by Eric Gravé.

silver staining, which makes the kinetosomes stand out—and, more recently, electron microscopy. Electron micrographs of the pellicle show it to be composed of hexagonal latticework, as if paramecium were held together by heavy chickenwire. Through each hexagonal opening, one cilium protrudes. At the base of each cilium below the pellicle lies the kinetosome; the rows of kinetosomes are called kinities, and running parallel to the kineties, and always on the right, are the kinetodesmatas.

The kinetosomes may divide to form new cilia, as they do when paramecium is preparing to divide, or they may, alternately and quite differently, divide to produce trichocysts. Trichocysts are dense, sharp-pointed bodies found in many ciliates and in a few flagellates as well; those of paramecium look somewhat like golf tees. They are discharged explosively through the pellicle when the animal is irritated. In some few protozoans they contain a toxin, or poison. Dileptus, for instance, carries a small arsenal of toxin-producing trichocysts around its mouth, which it uses both for offense and defense. In paramecium, however, their usefulness, if any, is obscure. Didinium, its mortal enemy, is totally unharmed by them, even though the threatened paramecium fires them by the hundreds. In fact, the trichocysts of most of the other protozoa as well seem to be completely harmless.

The method by which the trichocysts are discharged is also mysterious. In the discharged trichocysts, a long striated shaft can be seen; this shaft, which is attached to the blunt end of the little teelike object, is much longer than the trichocyst itself, yet it is not visible at all when the structure is still within the body of the animal. One current explanation is that the shaft is produced from some very compact, dry structure, which suddenly swells prodigiously—and in a split second—upon the addition of water, thus propelling the missile forth.

Paramecia of certain strains possess one curious set of weapons that *are* lethal, but only to certain other paramecia. These paramecia carry in their cytoplasm small bodies known as kappa (k for killer) particles, which they liberate into the medium. These kappa particles destroy other strains of paramecia, the "sensitives." Whether or not a paramecium is a killer depends on two separate factors. One is genetic; the capacity to harbor kappa particles is

inherited as a dominant gene. The other is cytoplasmic, the particles themselves. Even a paramecium that is a kappa by heredity
is not a killer unless it becomes infected by kappa particles from
some outside source. These then divide to provide the full quota,
which may range anywhere from two hundred to a thousand
separate particles. Kappa were among the very first bodies to be
found regularly in the cytoplasm of a cell that contained DNA. It
was first believed that the particles were small poison-producing
organelles; but now that they have been studied with the electron
microscope, most protozoologists agree that the particles were once
free-living bacteria-like organisms that settled down in a tolerant
strain of paramecium, setting up an enduring and mutually ad-
vantageous relationship. One is led to wonder, by analogy, if the
association between cells and other DNA-containing cytoplasmic
bodies, such as mitochondria and chloroplasts, might have come
about in the same way.

A PERITRICH

Vorticella represents another and quite different way of being a
holotrich. Vorticella looks more like a flower than an animal, an
impression confirmed by the fact that it spends most of its life
attached to the substratum by a long stalk. This stalk is transparent,
and within it can be seen a long contractile thread, the myoneme.
The myoneme is extremely elastic; the stalk can extend itself for a
great length, on a protozoan scale, or can suddenly coil up tight,
like a watch spring. Leeuwenhoek observed the movements with
great sympathy. "These little animals were the most wretched
creatures that I have ever seen," he wrote; "they . . . pulled their
body out into an oval and did struggle by strongly stretching themselves to get their tail loose; whereby their whole body then sprang
back towards the pellet of the tail, and their tails then coiled up
serpentwise, after the fashion of a copper or iron wire that, having
been wound close about a round stick, and then taken off, kept
all its winds. This motion of stretching out and pulling the tail
continues; and I have seen several hundred animalcules, caught
fast by one another in a few filaments, lying with the compass of a

FIGURE 46   Vorticella
stretches and contracts. These
movements are made possible
by fine fibers, the myonemes.
If you look closely, you can
see the myoneme within the
stalk.

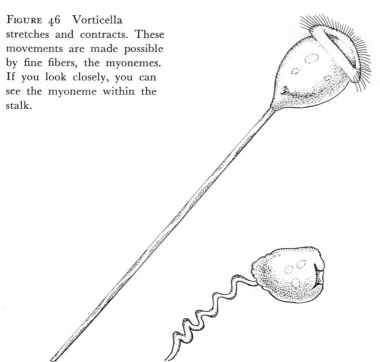

coarse grain of sand." The myonemes, the action of which Leeuwenhoek observed with such accuracy and sympathy, also continue
up into the cell, so that the whole body can contract. Some
scientists believe that they are similar in structure to the small
fibrils that make up our own muscle tissue.

On first impression, one is tempted to believe that vorticella is
open at the top, like a tulip, but actually, just where the tips of
the petals would be, there are three circles of cilia running counterclockwise into the funnel-shaped buccal cavity. It is this characteristic, the counterclockwise wreath of individual cilia, that identifies
the peritrichs, the group to which vorticella belongs. The beating of these cilia, as Leeuwenhoek observed, makes "a great bustle"
in the water and causes an inrush of food particles. Although—like
most peritrichs—vorticella spends nearly all of its life in one spot,
it can escape. Leeuwenhoek would have been less distressed had he

known that when the "wretched" animal becomes sufficiently annoyed it closes up its mouth, forms a little wreath of cilia around its posterior, abandons its stalk, and swims briskly away, tail first.

A SUCTORIAN

The suctorians are an aberrant group of holotrichs that possess cilia for only a brief portion of their life span. Tokophrya, which we have met before (page 62), is small and pear-shaped, attached to its substratum by a permanent stalk made of some nonliving material. In fact, after tokophrya dies, the stalk often remains. The adult suctorian has some fifty or sixty feeding tentacles, the remarkable structure of which was described in Chapter III. Tokophrya has attracted recent attention because of the rapid morphogenetic changes that take place in this tiny animal. Like the other suctorians, tokophrya reproduces by budding. First, a tiny opening, the birth pore, forms at the top of the animal between the two clusters of tentacles. Then, beneath the surface, below the pore, the embryo begins to form; it appears as if it is being scooped slowly out of the body of the mother animal as one might scoop

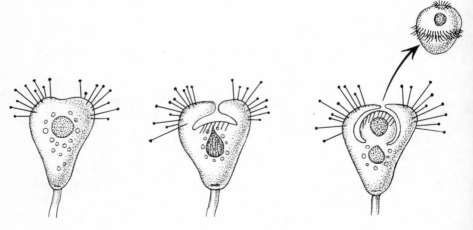

FIGURE 47 Budding in tokophrya.

out a melon ball. Cilia begin to appear—five rows of them—and even before the embryo has pinched itself off from the mother cell, the cilia start to beat; the kinetosomes of these cilia have migrated through the parent's body into the pre-embryonic birth pore to prepare for this event. Soon the larva is free of all connection with the adult cell but still trapped within the birth chamber. For ten or twenty minutes it whirls, beating its cilia, within the chamber. Then suddenly, like a bullet, it shoots out through the birth pore. The adult tokophrya is left empty and distorted, but soon composes itself and, in fact, within two hours or less may be ready to produce another embryo.

Once the larva escapes, there is a period of frenzied activity. The young potato-shaped tokophrya tears through the water with great speed in a highly characteristic whirling motion, zigzagging as it goes. This seemingly aimless activity may last from a few minutes to several hours. Then suddenly the larva stops and stands on what was formerly its head. It spins a few times more in place, the girdle of cilia beating about its equator. Next, right before one's eyes the creature transforms itself. The cilia, which were beating vigorously just a moment before, vanish. Then the attachment disc appears. No one knows with certainty where it comes from, but electron micrographs of its structure suggest that it may be formed by the coagulation of a droplet of some substance released from within the larva. The disc is very thin, composed of a network of fine fibrils, but it adheres so strongly that it will hold tokophrya fast even when several captured tetrahymena tug at its tentacles simultaneously.

Once the disc has formed, the stalk appears, and as it takes shape it lifts the animal up from the substrate. Next, and most remarkable, come the tentacles, moving out from the body, stiff as spokes, a dozen or so powerful, slender rods. All this happens almost simultaneously, and within five minutes the swimming embryo is transformed into a sessile form with all the characteristics of the parent organism. It is somewhat like seeing the frog change to the prince right on one's own doorstep.

Once it was thought that these tentacles might be created from kinetosomes borrowed from the body of the mother cell. Now that the electron microscope has revealed their fine structure as "seven-by-seven" instead of the now familiar "nine-plus-two," this no longer seems possible. Protozoologists are trying to capture enough of these little whirling embryos to pry them apart and examine them under the electron microscope in hopes of finding some clues as to how it all comes about.

CHAPTER X

THE CILIATES II
(THE SPIROTRICHS)

Stentor coeruleus, king of the ciliates, was described by one proto-zoologist as "without question one of the most beautiful animals in existence." It is imposingly large—a millimeter or more in length—and so can be seen by the unaided eye, although only as one might see a particle of dust caught by the sunlight. Its color, as the name implies, is a heavenly blue, and a rich culture of stentor will take on this same cerulean shade. The shape of the extended animal is most impressive. It is like that of a trumpet from which the genus came to be imaginatively named after "high-hearted, bronze-voiced Stentor," of *The Iliad,* "who could cry out in as great a voice as fifty other men."

Stentor is crowned with a wreath of membranelles. These are clumps of cilia, two or three rows thick and twenty to twenty-five cilia long, all adhering to one another. Each individual membranelle looks rather like a painter's brush; like such a brush, the roots are all bound together, forming a tight bundle. The bundle of each membranelle is connected through its root system to adjacent membranelles. They do not beat simultaneously but in a smooth rhythm, one following the other. These rows of membranelles, which are large and clearly visible even under low power, circle the head of stentor almost completely, beginning and ending at the funnel-shaped buccal cavity. Within this circular hedge of membranelles is a broad saucer-shaped depression, the peristome, which is carpeted with fine cilia. The beating of the membranelles,

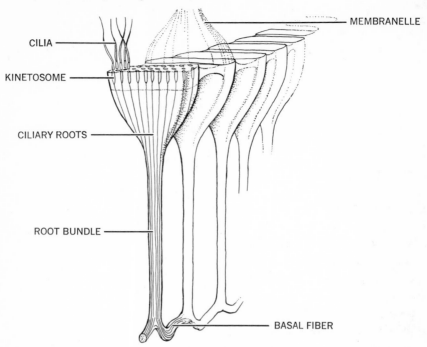

MEMBRANELLE

CILIA

KINETOSOME

CILIARY ROOTS

ROOT BUNDLE

BASAL FIBER

FIGURE 48 The structure of membranelles, showing how the cilia are gathered together in interconnecting bundles. From Tartar (after Randall and Jackson).

which always flows clockwise, creates a powerful vortex drawing a large and indiscriminate rush of particles into the peristome. The beating of the peristomal cilia is coordinated to move these particles toward the opening of the buccal cavity where they are spun around, some selected, and others rejected, apparently by a momentary reversal of the ciliary beat. Stentor eats bacteria, if necessary, but prefers larger entrees such as small rotifers and other protozoa, including, on occasion, stentor. The opening of the buccal cavity is able to extend to accommodate an outsize morsel, and, in fact, has been observed to close with such force as to hold a struggling rotifer or to bite off a large portion of a brother stentor. The funnel-shaped cavity, which corkscrews down to the cytostome can move in peristaltic waves when stentor is ingesting a particularly bulky meal.

The body of *Stentor coeruleus* is striped longitudinally; each animal has about one thousand stripes with clear and blue alternating, converging toward the apex, or tail pole. The cilia arise from kineties, rows of kinetosomes, underlying the clear stripes. These body cilia, which beat continuously, serve to move stentor through the water when it is swimming or to keep the currents moving, sweeping rejected particles out of the field and bringing in a fresh menu when stentor is attached and feeding. The cilia at the tail pole are slightly longer than the rest of the body cilia. When stentor decides to light, these cilia seem to seek out a favorable spot and effect a temporary attachment while the holdfast, as it is called, is being formed. The holdfast is made up of a group of stiff, sticky pseudopods at which stentor pulls as it stretches, like a balloon pulling at its anchoring ropes.

Underlying the kineties are slender contractile threads, the myonemes, similar to the myoneme visible in the contractile stalk of vorticella. These myonemes are highly elastic, permitting stentor to extend fully or contract into an almost perfect sphere. Stentor extended is from three to six times the length of stentor contracted. Stentor relaxed, which is the form in which it swims through the water, is an undistinguished cone-shape remarkable, nonetheless, for its color and for the conspicuous circle of membranelles that continue to beat and help to propel the animal through the water. Like almost all other free-swimming protozoa, stentor rotates as it swims.

Within the animal's large translucent body, the long nodular macronucleus is visible; the macronucleus of an average-sized stentor usually has about fifteen nodes, which are arranged regularly through its body. They must have some sort of attachment to the inner surface of the ectoplasm because they are fixed in position; they do not move when stentor rotates, nor are they dislodged by passing food vacuoles. The small micronuclei lie adjacent to the macronuclear nodes. Near stentor's anterior end, adjacent to the dark corkscrew-shaped gullet, is a large contractile vacuole, and near it is the cytopyge, which can be seen only when it is actually in use. The food vacuoles, which may be numerous,

137

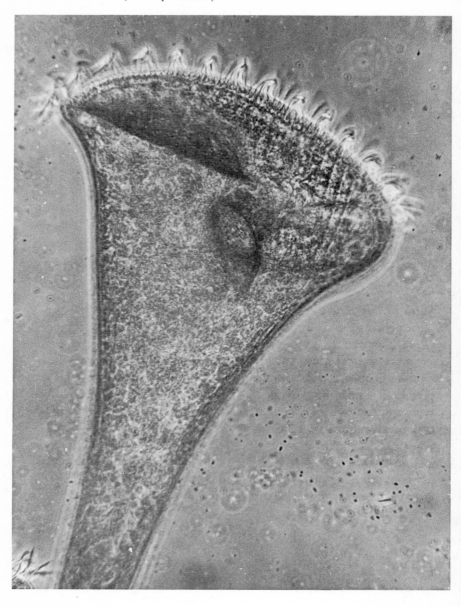

FIGURE 49 Head of stentor, king of the ciliates, showing its crown of membranelles. Photograph by Eric Gravé.

travel slowly through the body from cytostome down almost to the holdfast and back to the cytopyge in a regular procession.

THE HETEROTRICHS

Stentor belongs to a subdivision of the spirotrichs known as the heterotrichs. All spirotrichs, although they vary greatly in appearance, possess an elaborate system of membranelles running clockwise toward the gullet. Of the three major groups of spirotrichs, only the heterotrichs, however, are covered with fine body cilia as well as the larger cilia that combine to form the mouth parts. For reasons unknown, this group of ciliates are more likely to be pigmented than any others. One common type, blepharisma, for example, is a clear pink, while species of the stentor genus may be blue, rose, green, amethyst, yellow, or brown.

THE OLIGOTRICHS

The oligotrichs ("few hairs") are a group of spirotrichs that have no general body cilia. A good example of this group is halteria, which, as you may recall, made a brief appearance (page 25) in the chapter on locomotion. Halteria is almost completely bald, but its spherical body is studded with a girdle of clumps of stiff bristles, three or four in each clump and a total of seven to nine clumps, depending on the species. It jumps by giving these bristles a sharp flick. Around its mouth are a number—commonly fifteen—of large membranelles; within the buccal cavity are some ten smaller membranelles and, opposite them, a small undulating membrane. When halteria is not bouncing on its bristles, it rotates slowly in place by means of the clockwise beat of these powerful, seemingly outsized membranelles.

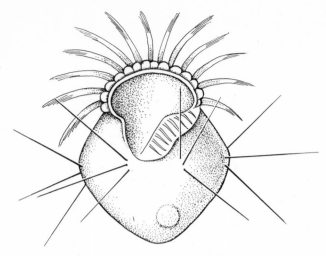

FIGURE 50 Halteria may spin in the water by beating its heavy wreath of membranelles, or alternatively, may bounce on its bristles. (50 microns.)

THE HYPOTRICHS

On December 26, 1678, Leeuwenhoek wrote, "This day there are some animalcules on which I can make out the paws and which are pleasant to behold because of their swift motions."

The little "paws," which are a characteristic of the hypotrichs, are actually cirri, tapering bundles of cilia pulled out into a point like the hairs of an artist's brush. The bases of the cirri are round, in contrast to the rectangular bases of the membranelles; otherwise, the structures are much the same. The cirri may be used as little paddles for swimming, aided by the prominent oral membranelles, which enable the hypotrichs to spin in spirals or circles or straight forward or straight backward. Usually, however, they are used for walking, which gives the hypotrichs a highly characteristic busy, jerky, scurrying motion. They have no other cilia except for some short, stiff bristles on their backs. Their pellicles are rigid, ranging in shape from almost round, as seen from above, to cucumber-shaped. In general, they are concave on the top and flat on the bottom, like beetles, and have a dozen or more "paws." Of the common types, euplotes is one of the most frequently studied. It has the advantage of having a transparent pellicle and cytoplasm, so that many of its structures can be seen at the same time.

Euplotes is commonly found in stagnant ponds and pools, even some that are slightly brackish; it also may turn up in a home aquarium, particularly one that is not tended too fastidiously. Euplotes is medium sized, about ninety microns long, smaller than paramecium, larger than tetrahymena. Viewed head-on, it can be seen to have half a dozen conspicuous longitudinal ridges, or furrows, running along its dorsal surface. Planted in small pits along these ridges are rows of stiff bristles. Seen from the side—and the patient microscopist can quite often find euplotes in the process of exploring a shred of leaf or other debris and so observe it from this vantage point—euplotes looks decidedly buglike, with its nimble, swiftly moving "legs." Clearly visible from the top, the most common viewpoint, are the large shadowy macronucleus; the single, fixed contractile vacuole, rhythmically pumping; and a procession of food vacuoles containing bacteria or very small flagellates. Through the transparent body it is also possible to see most of the cirri, although it is difficult to count them in the living, moving organism, and the conspicuous oral membranelles. Most of the ventral surface on the upper left side (euplotes' left, that is) is taken up by a large depression, sort of scoop-shaped, which leads into the buccal cavity. The row of membranelles, of which there are a total of between thirty-five and forty-five, begins on the right, dorsal surface of euplotes and sweeps down clockwise (as in all spirotrichs) toward the left, curving down onto the ventral surface to follow the left border of the peristome. Fine membranelles continue down into the buccal cavity, and opposite them, within the cavity, is the small undulating membrane.

All of these organelles—the eighteen cirri, the forty or so membranelles, and the undulating membrane—are coordinated in their motions. Whether euplotes is walking, paddling, or spiraling through the water, they move in synchrony. Furthermore, and even more difficult to explain, they can work one at a time; the cirri can remain motionless while the membranelles beat, or some

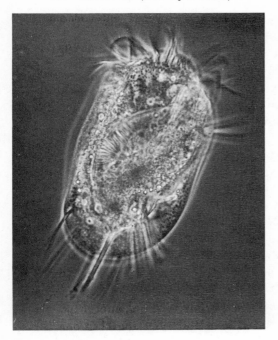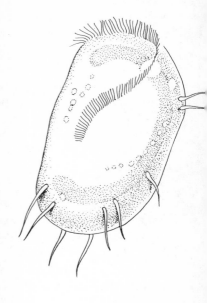

FIGURE 51   In this ventral view of euplotes, you can see the row of membranelles curving down around the cytostome, the large C-shaped macronucleus, and the "legs," or cirri. Photograph by Eric Gravé.

cirri can move while others are still. Fast and slow, forward and backward, these movements are negotiated with great facility. Microscopists have been able to trace within the body of euplotes a system of interconnecting fibrils, apparently linking the cirri with one another and with the membranelles, and converging near the cytostome. This converging bundle has been called the motorium, and many protozoologists have postulated that these interconnecting fibrils are a neuromotor apparatus, and the bundle, a sort of coordinating center or primitive brain. It has been shown that if one takes a glass needle and makes a cut in the side of euplotes, severing the fibrils leading from the caudal cirri to the motorium, coordination between the movements of these powerful "paws" and the rest of the organelles is lost. A similar cut elsewhere in the body of the animal does not produce a similar effect. One still does not know, however, what other, perhaps still invisible, structures have been severed or what the general effect of such an onslaught might be.

Euplotes is small; all of these activities and intricately moving parts are condensed in an area less than 1/250 inch. Yet, once you have watched euplotes, if only for a minute or two, you may find it difficult ever again to dismiss the protozoa as the "simplest of all animals," or perhaps even ever again to speak of any cell as simple, so extraordinary are the potentials that may be hidden in the "primitive living jelly" of its interior. It is as if Nature, to use that old-fashioned term, asked herself how much she could get into one single cell. And euplotes and its fellow hypotrichs are the answer.

CHAPTER XI

THE PARASITIC PROTOZOA

The life of a protozoan is filled with perils—cold, drought, hunger, and a host of predators haunt its microcosm. One particularly satisfactory way for a one-celled animal to solve these problems is parasitism. Thousands of species have taken up this way of life, living either *on* or, even more conveniently, *in* some other, larger animal, achieving for themselves and their progeny free transportation, an abundant food supply, and a sheltered environment. The parasitic protozoa are found everywhere. One group, the thigmotrichs (a type of ciliate), specializes in clams, mussels, and other mollusks, clinging to their gills or mantle cavities by a special little tuft of cilia. Some thigmotrichs merely enjoy the bountiful food supply which flows constantly past them by courtesy of their host. Others drive powerful little suckers into the cells of the gills or mantle and drain out the cytoplasm. The former belong to a group known as "commensals," which means "eating at the same table"; the latter are true parasites, which live at the host's expense.

Other common commensals include a group of ciliates that specialize in sea urchins, and opalina, a small leaf-shaped animal covered with oblique rows of cilia, which lives in the intestinal tract of frogs. A single sea urchin has been found to harbor as many as twelve different species within its interior, and a frog without one or more of the two-hundred-odd species of opalina is a biological rarity. Neither the sea urchin nor the frog seems to suffer from their hospitality.

Another group of marine ciliates, the apostomes (animals with very small mouths), are found largely on crustacea. One is a parasite of the hermit crab, with which it lives in a strange relationship. The ciliate encysts on the crab gills and then, at the time when the crabs molt, hastily excysts and penetrates the crab's shell, where it rapidly divides. When molting is completed, the ciliate parasites, now greatly increased in number, are carried off on the discarded exoskeleton. They feed on the castoff shell, increasing their volume up to thirty-two times in six to ten hours, escape from the exoskeleton, and then divide. They may enjoy a free life for six to eight days until fortune brings them another crab, on whose gills they promptly encyst until molting time. Apparently the chances of finding a crab and the reproduction rate of these parasites are in balance, so that the population neither dies out nor increases.

Not only are the "lower animals" hosts to parasites but so are all species of the vertebrates. Man, too, harbors many varieties of protozoa, ranging from quite harmless amoebas inhabiting his digestive tract to the trypanosomes, the flagellates that cause the often deadly African sleeping sickness. Most of the parasitic protozoa of vertebrates are inhabitants of the digestive tract, and those that are not, such as the malaria parasite, seem to be the evolutionary descendants of such parasites.

PROBLEMS OF PARASITISM

One major problem a parasite, whether of man or mollusk, has to face is the extent of damage it may do to its host. An example of such "overparasitism" is the myxoma virus of rabbits that was introduced into Australia to destroy the rabbit population (which man had previously brought to Australia some one hundred years before). At first the myxoma virus was so deadly that the coneys often died too rapidly to have a chance to infect another rabbit; so the virus died too.

Thus, the deadliest strains failed to survive and were eliminated —the virus became progressively less virulent. At the same time, natural selection was working on the rabbit population and the rabbits became less susceptible to the virus than their predecessors. Now, as a consequence, some 95 percent of the rabbits survive infection by the myxoma virus.

Examples such as these lead many parasitologists to argue that parasites that harm their hosts are probably the result of recent evolutionary relationships—the two simply have not "learned" to get along well together yet. This view is supported by the fact that parasites that alternate between two different types of hosts are usually harmful only to the more recent arrival on the evolutionary scene. For instance, the many parasites that travel from vertebrate to vertebrate by way of insects, multiplying in both hosts, often harm the vertebrates but never the insects.

On the other hand, it is equally logical to argue that most parasites first made their way into the host, evolutionarily speaking, quite by accident. This point of view is supported by the fact that the intestinal tract is the most common abode and that many parasites residing in the gut are closely allied to free-living forms. The invasion of the central nervous system by the flagellate that causes African sleeping sickness or of the red blood cell by the malaria plasmodium must represent the end product of millenniums of increasing encroachment.

A third host-guest relationship, and one of particular interest, is symbiosis—"living together"—in which, at the very least, each organism benefits from the mutual relationship and, at the very farthest extreme, each cannot survive without the other. It is possible that certain structures now considered organs of a host animal once began as symbionts, such as kappa (see pages 129–30) and perhaps—some biologists have hypothesized—even organelles such as the mitochondria and the chloroplasts.

Among the most successful symbionts are dinoflagellates, which are found in a number of marine animals, including jellyfish, sea anemones, and coral. The body of the animal offers the small dinoflagellate a sheltered environment and, perhaps, a source of carbon dioxide for photosynthesis. The dinoflagellates, in turn, appear to produce nutrient materials or growth-stimulating factors —perhaps vitamins—for their host. Coral, for example, usually contains a large number of symbiotic dinoflagellates, and laboratory experiments have shown that coral grows ten times faster in the presence of light—when photosynthesis can take place in the symbiont—than in darkness. In nature, the actively growing area of a coral reef is close to the water surface, where photosynthesis can take place, although the corals themselves are carnivorous. In their symbiotic form, these dinoflagellates are always in the palmellar stage. In order to distinguish them from simple algae, much less to identify their species, it is necessary for the marine biologist to separate them from their host and coax them into a free-swimming form.

TRANSMISSION OF PARASITES

Besides the question of maintaining a delicately balanced relationship with its host, the parasite also faces the problem of providing for the future of its race. Survival, in biological terms, has little to do with the survival of the individual but a great deal to do with the provisions the individual makes for its progeny. Since the life span of any one host is limited in time, the parasite, to ensure perpetuation of the species, must have a way of traveling from host to host. This is sometimes easy. Ciliates that live in sea urchins have no problem at all; sea urchins are commonly given to cannibalism, so their protozoa enjoy a free ride from host to host. Opalina, the pretty commensal of the frog, also has a relatively simple system. It multiplies with great rapidity as the weather gets warm in the springtime, so that when each new crop of tadpoles appears— always an annual springtime event—they are born into water generously seeded with encysted opalina, which they ingest.

The intestinal amoebas of man pass from host to host in a similar fashion. They cannot nourish themselves outside of the intestinal tract, so accustomed have they become to this particular cultural milieu; before that they can only survive in the outside world in the form of cysts, and it is these cysts that find their way into new human intestines. Most of these amoebas are harmless, but there is one, *Entamoeba histolytica*, which can become particularly troublesome. This amoeba usually lives as a commensal in the large intestine, where it feeds on bacteria and other food materials that the intestine contains. Sometimes, and for reasons quite unknown, it abandons this commensal existence, attaches itself to the lining of the intestine, secretes an enzyme that eats its way through the cell lining so that the amoeba is able to invade the intestinal walls, which it then spreads along, causing the ulcers that are characteristic of amoebic dysentery. The infective stage of *Entamoeba histolytica* is a special cyst with four nuclei that is formed in the lower intestinal tract. When this protective cyst is ingested by another human being, it passes through the stomach into the small intestine, where the digestive enzymes of the host dissolve away its tough, resistant coating to liberate the quadrinucleated amoeba. The nuclei then divide to form eight new nuclei, and then the cytoplasm divides to form eight new amoebas, by now all comfortably lodged in the large bowel.

Other protozoan parasites lead more complicated existences, having chanced upon a method of traveling from host to host that depends on an intermediary host, which serves as a vehicle—or, as it is technically known, a vector. The trypanosome, the flagellate that causes African sleeping sickness, is a protozoan of this type. Its vector is the tsetse fly which, as it bites its victim, deposits the flagellate into the blood stream, where it multiplies and, usually after a matter of weeks or months, makes its way into the cerebrospinal fluid causing torpor, and finally, if the victim is untreated, death. Usually by this time—since the disease is protracted—the patient has been bitten by another tsetse fly, which then carries off a sample of flagellate with which to start a new colony.

Leishmania donovani, also a flagellate, employs the sand flea to make its way from host to host; in the insect, it is harmless, but in the human it is the cause of a distressing disease of the internal organs, known as kala azar or leishmaniasis, widespread throughout Africa and the Far East. Insects and human beings offer very different environments; as a result, the protozoan has to become almost two different animals to accommodate itself to each. Leishmania, for instance, within the human tissue cell, is small and ovate; in the sand flea it is an elongated flagellate, with one anterior flagellum and another which extends backward but which, rather than moving freely, lies under the animal's pellicle. Such a flagellum, which vibrates as the flagellate moves, is known as an undulating membrane—a somewhat confusing appellation since it bears no relationship to the membranelles of the ciliates.

One must remember that to call the tsetse fly or the sand flea a vector is a particularly anthropocentric point of view. If the insects or the parasites could be consulted on such a question, they might well consider that man is a peculiar species provided by evolution for moving parasitic protozoa from one insect to another.

THE SPOROZOA

The rhizopods, flagellates, and ciliates all have their parasitic forms. In addition, there is another major group, the sporozoa, all of whose members are parasites, in the full sense of the word. Sporozoa is also a confusing designation because "spore" is one of those words that biologists use to mean just what they choose it to mean. A spore may mean a resistant body formed by a bacterium to weather difficult conditions, or the asexual reproductive cell of a plant, depending on whether a bacteriologist, or a botanist is speaking. To the protozoologist, a spore means the daughter cell that results from multiple fission following some sort of sexual phenomenon, and all the sporozoa typically reproduce by sporulation, although many reproduce asexually as well.

The gregarines, for instance, are a group of very common sporozoa that parasitize insects. In most species of gregarine, there is no asexual reproduction. Two gregarines of different "sexes" encounter one another in the intestinal tract of their common host. The pair then becomes encysted in a protective envelope, the gametocyst. Then each member of the encysted pair divides to form gametes, which mate with one another, each always choosing an offspring of the opposite parent. The zygotes (the result of fusion of the gametes) encyst in turn, still within the gametocyst, and then divide three times. Here they often actually enter one of the cells lining the intestine, curl up in it, and live there until they become too large. Even after they emerge, they often remain fastened to the cell by a special mouth part called the epimerite. The epimerite may take many forms, ranging from a simple knob to a quite elaborately structured sucker or a cluster of rootlets. Once thoroughly fed, the gregarine loses its hold on the now emptied cell and wanders through the intestinal tract. During this stage, gregarines are often found strung together in long, gregarious

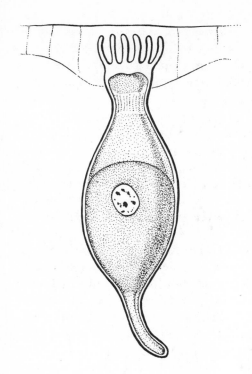

FIGURE 52 The parasitic gregarine remains attached to the cell of its animal host by means of special mouth parts. (Up to one millimeter.)

chains, end to end; this association is known as syzygy (a term of great potential for crossword-puzzle fans). Eventually, however, they give up this social life, choose a partner, and begin the cycle anew. The gregarines, although they are ubiquitous, do little damage to their hosts because, first, they do not multiply inside the host, so their numbers never become overwhelming, and, second, each gregarine destroys only one cell—a cell of a type that normally multiplies rapidly, so the damage is repaired quickly.

On the other hand, the coccidia (a group of sporozoa common among vertebrates) do multiply in the host, sometimes prodigiously, and are the cause of serious diseases among fowl, rabbits, pigs, cattle, sheep, and almost all other domesticated animals—which makes them a great economic hazard—as well as among wild vertebrates. As with the gregarines, the young coccidian enters an intestinal cell of its host, and goes through another process found typically among the sporozoa, schizogony. In schizogony, the well-fed coccidian cell undergoes a number of different nuclear divisions, all of which take place within the original cell membrane. Then, when the nuclear divisions are completed, membranes are laid down within the cell, dividing the cytoplasm into equal portions, each containing a nucleus—precisely as one would divide up a flat square of cake. In one species of coccidian that infects cattle, one single cell is characteristically divided up into an astonishing total of one hundred thousand new cells. These new cells, merozoites, as they are called, then penetrate other intestinal cells, once more multiplying a hundred thousand fold. Occasionally merozoites, however, differentiate sexually. The female-to-be enters a host cell and there grows to an extremely large size without undergoing any nuclear division or schizogony, simply storing a food supply for the future generation. The male-to-be similarly enters a host cell and there divides, not into merozoites but into many flagellated microgametes. The male and female cells find one another in the intestinal tract, mate, and immediately the zygote forms a tough resistant cyst, which is eliminated from the intestinal tract. This cyst is the infective form, entering a new host that shares the same barnyard, grazing land, or water supply.

The most familiar and most thoroughly studied of the sporozoa is plasmodium, the parasite that causes malaria. Plasmodium has probably done more to influence the course of human history than all the monarchs, ideologies, explorations, and human aspirations combined. Its life cycle is complex. A female anopheles mosquito bites a patient with malaria and withdraws through her slim hypodermic proboscis the blood she requires to nourish herself and her eggs—often she takes more than her own weight in blood—and also, all too frequently, a sample of plasmodium. (The male mosquito, having no eggs to nourish, is quite content with a more sybaritic diet of nectar and fruit juice and does not even have the needlenose of the female.) The formation of spores takes place in the stomach of the mosquito, following which the tiny spores break loose and make their way to the salivary glands. This takes about a week or ten days and, since mosquitoes have an average life expectancy of about three weeks, it represents a very tenuous phase of the existence of this particular tribe of plasmodium. After this time, ideally (from the parasite's point of view), anopheles finds another victim; just before she bites into the capillary of the victim, she injects a droplet of spore-laden salivary fluid into the skin. This fluid is an anticoagulant, insuring smooth passage of the thin stream of blood from man to insect, and is also thought by some to have anesthetic qualities (accounting for the fact that we almost never feel a mosquito bite until it is too late). The all-too-familiar welt left by the mosquito is the skin's reaction to this droplet of salivary fluid.

Once the slim, sliver-shaped plasmodium spores get inside the human blood stream, they go into hiding for a brief period in the cells that line the capillaries of the liver and perhaps in other tissues as well. This is known as the hidden, or "crypto," stage, because during this period the patient has no symptoms of the disease. In the liver, the animals divide by schizogony, releasing a host of tiny round merozoites. Some of these merozoites go back into hiding in the tissue cells; it is this fifth-column contingent of the plasmodium that, emerging from hiding from time to time, is responsible for the recurrent attacks so characteristic of malaria. The rest of the merozoites—the majority of them—invade the red cells circulating in the blood stream. Within the red cell, the merozoite becomes almost amoeboid, developing pseudopodia and pinching off portions of the hemoglobin of the red cell and phagocytizing them. It finally grows until it occupies most of the blood cell. Nuclei form (from eight to twelve depending on the species), membranes are laid down within the cytoplasm surrounding them, and the mass divides to form a whole group of new merozoites, which burst from the ravaged cell. This same growth cycle has been taking place simultaneously and synchronously in all the infected blood cells of the patient, and it is at this moment, largely from the release of the toxic debris of the destroyed cells, that the typical chills and fever of malaria occur. In fact, because of the different rates at which the two most common species divide, a physician can tell exactly which type of malaria his patient has, simply by observing whether his patient develops a high fever every forty-eight hours or every seventy-two hours. Once released, the merozoites invade fresh red blood cells to begin the cycle anew.

Some of the merozoites, although they are indistinguishable from the rest, do not undergo schizogony in the red blood cell. Rather they develop into gametes—large amoeba-like organisms with a single nucleus, some female (the larger ones) and some male. This is the way that plasmodium provides for the future of its race. These gametes remain in the human red blood cells waiting for the female anopheles to rescue them. If this happens—which is not too unlikely in a tropical climate—the imprisoned male and female gametes are released from the human blood cells into the stomach of the mosquito. The male gametes form some four to eight filament-like structures, one of which fertilizes a female gamete. The fertilized egg divides by multiple fission within the stomach of the mosquito and the sporozoa break loose, traveling to the salivary gland, where they are ready for microinjection into the next human being. Plasmodium does not seem to harm the mosquito vector in any way.

This curious life cycle, embracing man, malaria and mosquito, has clearly been in existence as long as recorded time and probably for millions of years more, but one can only speculate as to the chain of events that brought the three species together and adapted them so well to the service of protozoan parasites.

TRICHONYMPHA

If you are ever afflicted with termites, you will at least have a chance to glimpse a host-parasite relationship that is one of the most highly specialized. Packed absolutely solid in the gut of the termite—and in fact accounting for as much as one third its total weight—is a collection of extraordinarily beautiful one-celled animals. The termite and its protozoa are a perfect example of symbiosis; neither can survive without the other, and together they constitute, as many householders have learned to their sorrow, a biological entity of extraordinary efficiency.

The most frequently encountered of these symbiotic protozoa, and also the largest and handsomest, is *Trichonympha campanula*. This species of trichonympha, as its name implies, is shaped like a bell—or, actually, is more the stylized bell shape of some intricately fashioned Christmas tree ornament. Most of its body is

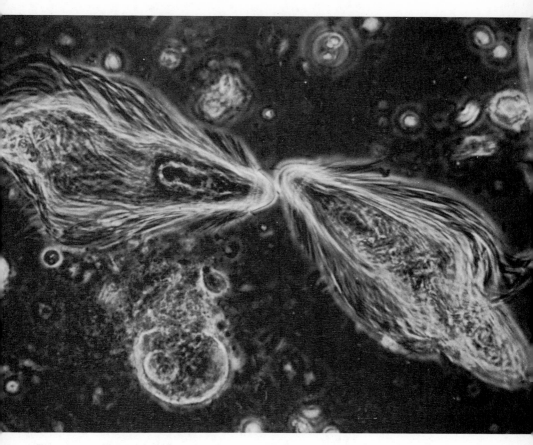

Figure 53 Trichonympha, showing the shimmering white flagella with which the animal is covered. (Up to 300 microns.) Photograph by Eric Gravé.

covered with long, fine flagella, shimmering like tinsel. When trichonympha is released from its host—a simple surgical procedure —these flagella can be seen to beat quite actively, although not with the smooth coordination of the ciliated protozoa, and its protozoan body writhes in response to the flagellar motion. There seems to be no reason for trichonympha to have so many flagella and to be capable of such vigorous activity. Clearly, in trichonympha's sole abode such motion is unnecessary and probably even impossible, so close-packed are the cellular inhabitants.

The thousands of shimmering flagella conceal a quite complicated architecture. The body of trichonympha is divided into three distinct parts. The anterior section, known as the rostrum, is mounted by a small apical cap, a tiny bald dome, from below which flow the longest of the flagella. These flagella ordinarily conceal a deep circular groove, or fissure, which virtually separates the rostrum from the midsection of the body. This groove is so deep that the two parts of the animal are held together only by a slim tube of endoplasm; sometimes as the animal writhes, one catches a glimpse through the flagella of this tough, flexible core, or "neck." The midsection is also luxuriantly flagellated. It broadens toward the posterior, giving the animal its characteristic bell shape. The flagella farthest from the anterior tip are the longest and they hang down over the soft bare posterior third of trichonympha like a skirt. From the apical cap through this midsection, trichonympha is striated with deep longitudinal furrows; the flagella emerge from the body of the animal through these longitudinal grooves.

Trichonympha's posterior is sticky and amoeboid. Wood particles eaten by the termite cling to the animal's soft "belly" and are infolded amoeba-fashion and digested, and the juices are emitted to nourish the termite. The termite can be "cured" of trichonympha simply by withholding food from parasite and host for a week or so or incubating them at a high temperature, in which case the protozoan is the first to succumb. Cured termites continue to eat, but to no avail; they completely lack the enzymes necessary to digest the tough cellulose of the wood fibers they take in, and soon die of starvation, stuffed with useless nourishment. Also, if young termite larvae are removed from the nest before they can be infected by the droppings of the adults, they do not survive.

Trichonympha, similarly, is completely dependent upon its host. No protozoologist has been able to devise a medium in which trichonympha can live, as it does so readily in the viscous interior of the insect. In the wood roach, with which trichonympha also lives a symbiotic relationship, the sexual reproduction of the protozoan is entirely governed by the host. Trichonympha usually reproduces asexually, by cell division, forming two mirror-images in conventional flagellate fashion. It also has a sexual cycle. Simultaneously all of the animals infecting a particular roach form cysts around themselves, and their nuclei divide. The nucleus of trichonympha, which is located just below the rostrum, is large, and its twenty-four chromosomes can be clearly seen when the dividing cell is treated with special stains. Within the cyst two gametes are formed, one male and one female. Then the cysts break open—all of them—releasing the gametes into the termite gut. The gametes originally resemble one another, but soon the female develops a fluid-filled bubble at her posterior extremity, and the male gamete penetrates the cell through this bubble and then completely dissolves. The last structure to remain visible is the rostrum, which looks like the small, discarded nosecone of a miniature spacecraft. The male nucleus fuses with that of the female cell. The fertilized cell now has double its normal quota of genetic material and so quickly undergoes a series of reduction divisions in which new individuals are produced, each of which possesses the normal complement of twenty-four chromosomes. This remarkable and well-synchronized series of events takes place during the periodic molting of the roach. And it has been found that the hormone that causes the roach to molt also acts directly on trichonympha to stimulate its sexual cycle, thus to assure that all the gametes are formed simultaneously so that all may find a mate. The sexual cycles of several other protozoan inhabitants of the roach are similarly triggered by the molting periods, but each occurs at a different and absolutely characteristic stage in the molt.

Some thirty-five years ago, a particularly unusual symbiont was found among termites in Australia. This animal, which is large and somewhat resembles trichonympha in general body contour, was given the name *Mixotricha paradoxa*, because it appeared to contain both cilia and flagella. It has four relatively short flagella protruding forward, and its body is covered with what first appeared to be cilia arranged in neat oblique rows over its surface. Recent studies, aided by the electron microscope, have shown that these are not cilia at all, but bacterial cells, spirochetes. The pellicle of *Mixotricha paradoxa* is covered with evenly spaced, characteristically shaped knobs. Each of these knobs serves as a bracket to which one threadlike spirochete is attached. These spirochetes all

move in unison and their undulating movements propel the protozoan, while its own flagella serve merely as a steering apparatus.

In these particular situations, the words "host" and "parasite"
have lost meaning; the two animals—the wood roach and trichonympha, and the protozoa and the bacteria—have become single
organisms.

CHAPTER XII

THE COLONIAL PROTOZOA

At some unrecorded point in time, perhaps very early in the history of the protozoa, certain of the one-celled animals began to form societies of cells—communities in which different cells had different functions—and from these societies the metazoa, the many-celled animals, are believed to have evolved. Fossil traces of early metazoa that date back over half a billion years have been found, but the soft, small bodies of the earliest metazoa have probably left no imprint in the mud into which they fell, and those who wish to read this chapter of evolution must seek its traces in the animals that are alive today.

Since the course by which the metazoa came into being will probably never be known with certainty, this subject has long formed a rich field for the play of the imagination. Some have speculated that perhaps a multinucleated ciliate laid down additional membranes to form many cells within its single-celled body. Others believe that perhaps groups of different kinds of cells came to live together in symbiosis, just as the bacteria-like kappa particles (pages 129–30) have moved in with paramecium, or as the palmella dinoflagellates have taken up residence in the radiolarians and the corals. Eventually each symbiont might have so surrendered its autonomy so that the separate cells merged into a single organism of specialized cells, each with limited functions.

The most widely held theory, however, and the one supported by the most evidence, is that the multicelled animals arose from colonies of cells, which formed much in the same way that the embryo is formed from the egg cell, by repeated cell divisions that produce cells that remain together rather than become separate units. Eventually these cells begin to take on special functions and lose others until the group becomes interdependent, the colonies becoming true societies.

Almost any type of protozoan cell might theoretically have given rise to such a colony; indeed, perhaps different types of metazoa have originated in different types of single cells. There does seem to be good evidence, for instance, as mentioned on page 106, that the porifera (the sponges) evolved from one special type of chrysomonad. It seems most likely that the sponges are a group apart, however.

A number of biologists believe that both modern animals and modern plants had a common origin in small green flagellated cells. Such cells, because they are less highly structured than many other protozoa, can lend themselves more readily to colonial forms, having more reasons to be cooperative and less to lose in the way of intricate organelles and appendages. Furthermore, the sexual reproduction of such small green cells as chlamydomonas appear to offer another clue to metazoan evolution. One of the specialized cells most urgently needed by a complicated many-celled animal is a reproductive cell, and the existence among the small green flagellates of flagellated gametes not too unlike those of many of the higher animals, including ourselves, seems to weigh the evolutionary argument in favor of this cell type. It is of interest that ferns, among the most ancient in heritage of the higher plants, also have swimming flagellated sperm.

The question of why the metazoa came into existence is a little clearer. On the one hand, the great success of the protozoa is without question. The greatest advantage to being one-celled is probably prolificacy; a single cell can replicate very rapidly, often several times a day, and produce instant adults, avoiding all the vulnerabilities of immaturity. A second advantage, and one related to this extraordinary reproductive capacity, is the great variety of forms and functions of the single-celled animals, a variety that permits their survival where no other living things are found.

There are disadvantages too, however, and the chief one is probably size. The larger the animal, generally speaking, the faster and more mobile and less edible it is, particularly at the protozoan end of the food chain. Some of the protozoa have become very large, and indeed common forms like stentor are larger than many of the "simpler" of the metazoa, such as rotifers and the smallest worms. But the rotifers and worms have chosen a "wiser" course from some points of view, for there is a stringent limit to the size of a one-celled animal. This limit is imposed chiefly by the relationships between volume and surface of a single cell. A ping-pong ball, for instance, has far more surface in relation to its volume than does a baseball, and a baseball has far more than a basketball. In living things, as distinct from basketballs, the question of surface is crucial because cells, *all* cells, carry out a great number of their vital processes—excretion, respiration, and pressure regulation—at their outer membrane. The individual cells of the mouse and those of the elephant, for instance, are just about the same size, indicating that there is an optimum size for these surface functions, and these mammalian cells are in general much smaller than those of the protozoa. Evolution has provided some of the larger protozoa with more membrane surface. This is achieved by the development of unusual cell shapes—long and tubular, or broad and flat. Many also have pellicles with pleats, or ridges, or can reach out massive pseudopods to increase their area of contact with the environment. None of these forms or changes, however, provides as useful a solution to the problem of surface-to-volume ratio as the multicellular state, in which each small cellular unit is in direct contact with the external or internal environment.

PALMELLAR STAGES

Many of the green flagellates secrete for themselves a layer of clear gelatinous material, which coats their outer membrane or their cell wall. This is generally invisible under the microscope unless a microdrop of India ink is added to the slide to stain the medium. Under these circumstances, the jelly-like outer layer shows up as a clear halo surrounding the cell. Some types, notably chlamydomonas, euglena, and certain dinoflagellates, go through stages in their life cycles in which they surround themselves with quite large amounts of this same sort of mucilaginous material, discard their flagella, and embed themselves in this sticky matrix— bright specks of green in a clear jelly. In this gelatin mass, which protects them from the full force of the rays of the sun and perhaps also from some predators, the cells divide. Chlamydomonas is usually found in groups of gelatin capsules each containing two or four daughter cells. Euglena often divides five or six times, so that each gelatin mass frequently contains thirty-two or sixty-four cells. Finally, division completed, the cells grow flagella, escape from the mucilage, and swim off.

Another way in which cells form casual aggregates is simply by not separating completely after cell division. Many of the non-motile algae and the fungi form long filamentous ribbons of cells, in which each cell exists entirely as an individual. Similarly, some of the flagellates stay clumped together after cell division, although no cell takes on any specialized role and there are no cytoplasmic connections between them. This is seen particularly often among cells that exist by photosynthesis or by absorbing soluble food through their membranes. Cells with these modes of nutrition are more readily able to live in close company than types that must hunt and forage.

One common flagellate that leads a community existence is gonium, although as a society gonium can only be regarded by even its warmest admirers as a model of poor planning and mis-management. Each colony is made up of four to sixty-four cells depending on the species. These cells are arranged in a flat disc or shield, with some of the cells forming a square in the center and the others forming a border around the square. The individual cells are small—only about ten microns in diameter—green, ovate, and cellulose-walled, like chlamydomonas. Their flagella are all pointed in the same direction, which is at right angles to the surface of the shield. Each beats its flagella frantically, striving to swim like a free-living flagellate, and as a result the entire shield is pulled forward in such a way as to offer the absolute maximum in water resistance. From time to time the colony seems to lose its balance entirely, tumbles, somersaults clumsily once or twice, and then rights itself, pushing valiantly on.

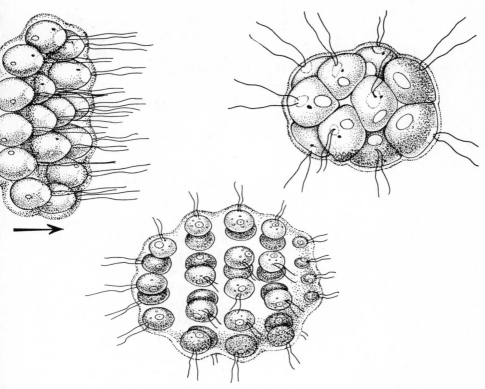

FIGURE 54   Three cellular colonies, gonium, pandorina, and pleodorina. Gonium is the shield-shaped colony (left, 90 microns). Pandorina (right, 50 microns) and pleodorina (center, up to 450 microns) are roughly spherical and so are able to move more efficiently than gonium.

A closely related colony is pandorina, which forms a close-packed rather lumpy oval mass of eight, sixteen, or thirty-two cells embedded in a clear mucilage; each cell has two flagella, and all the flagella point outward, so pandorina tumbles through the water like a ball. When the cells attain their maximum size, the colony sinks to the bottom and each one of the cells divides four times, forming a subcolony of sixteen cells. The subcolonies remain together until each one is completed, and then the parent breaks open, like Pandora's box (from which it gets its name), releasing sixteen new daughter colonies, each containing sixteen cells.

Pleodorina is also made up of tiny green flagellates embedded in a jelly sphere. It differs from gonium and pandorina in one interesting particular; certain of its cells, usually less than half, are incapable of reproducing to form new colonies. These cells are usually smaller than the others. This giving up of one vital function is actually an evolutionary step forward.

ZOOTHAMNIUM

Ciliates also form colonies, some of which are very precise and well ordered. One such ciliate is zoothamnium, a peritrich related to the bell-shaped vorticella. There are several species, both marine and fresh-water, all of which take on an elegant treelike shape. They differ from other stalked ciliates not only in their precise relationship to one another but also in the fact that they are interconnected by a thread of protoplasm running through the stalk. If one of the ciliates is touched, even slightly, either the whole branch or the entire colony, depending on the species, will contract right down to the base. The irritability of the colony and its delicacy make it particularly frustrating and difficult to study. Even the touch of a cover slip can serve to make the entire colony disappear from the microscopic field.

One species, *Zoothamnium alternans,* has been examined in particular detail. Each new colony begins from a single cell in a previous colony which enlarges, becomes somewhat barrel-shaped, loses its aboral cilia, develops a ciliary girdle, breaks off from the family tree, and swims away. Such a cell is called a ciliospore. The ciliospore leads a migratory existence for only a few hours in its entire life span. Once it has found a likely spot—and it may first try two or three—it settles down, forms a base, and then begins to build a hollow stalk. After it rises a short distance off this stalk, it begins to secrete into the lining of this hollow tube the protoplasmic thread that serves as the easily triggered contractile muscle. By this time, the ciliospore has resumed its original shape.

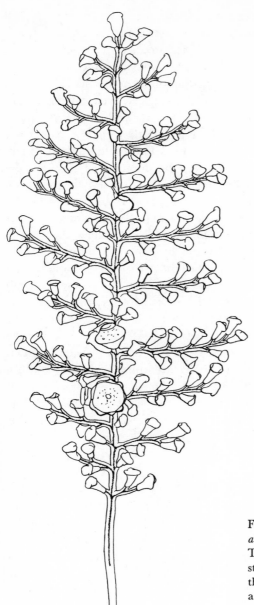

FIGURE 55 *Zoothamnium alternans,* a colony of ciliates. The larger cells close to the stem are reproductive cells that will break loose to form a new colony. Colonies range in size up to several millimeters. After Summers.

After the protozoan has raised itself about three or four cell heights off of its substrate—which is commonly a turtle's shell—the first cell division takes place. This produces a larger cell that remains at the tip, or apex, of the stalk and a smaller cell on the side of the stalk. Each cell in turn produces more stalk, lining it with protoplasmic thread. The top, apical, cell then divides again, producing once more a larger cell at the uppermost tip and a smaller cell on the side, but this time, on the opposite side from the first cell (hence the name *alternans*). The zoothamnium colony increases in height, by division of the larger cell (called the macrozooid); the smaller daughter cells that are left behind, each on alternate sides of the main stalk, also begin to divide. These too follow a pattern like those of the apical cell; each one divides, giving rise to a lateral daughter cell, somewhat smaller (the branch cell), and the cell at the tip of the branch. The lateral branch cells never divide, only the cells at the tip of the branch; and even they are limited in their division. Only a few branch cells are permitted on the first branch; more are permitted on the second and third, and so forth, so that the entire colony curves gracefully out at the center and then back in again at the tip. As many as thirty-three generations, or branches, have been counted in a single colony, and twenty or so are quite common. Within each colony, some cells develop into ciliospores. Only a few of these, however, seem to be candidates for colony formation. The most likely seems to be the first branch cell on the very first, or bottom-most, branch of the colony. Other cells may also sometimes become ciliospores, but these, again, must be branch cells right next to the main stem.

Early workers hypothesized that at the time of division each daughter zoothamnium cell received a different endowment so that it was genetically ordained that one should be a nondividing branch cell and another should continue to produce stalk and daughters. A series of delicate experiments and careful observations have shown that this is not true. If one or even several cells at the end of a branch are nicked off, the next remaining cell will take on the function of its erstwhile neighbor and begin to divide. Similarly, if the apical cell is severed from the colony, the first branch cell on the uppermost branch will assume its function.

There is, however, one restriction: Branch cells are able to grow and divide only if they have not been branch cells for too long. After a time this potential seems to disappear.

These ingenious experiments with *Zoothamnium alternans* recall experiments done some years ago with frogs' eggs. These are among the most common and best-studied tools of the embryologist, who over the years has been able to construct so-called "fate maps" of the developing egg. These maps, which are prepared by marking various regions of the egg or embryo with harmless vegetable dyes, made it possible to predict from which identical cells the various highly differentiated tissues of the animal are "fated" to develop. In the 1920s the famous German embryologist Hans Spemann showed that if cells were removed from one place and replanted in another in the early stages of embryo growth, those presumably destined to be one sort of tissue, such as skin, would differentiate into another, depending entirely upon their new location in relation to other cells in the developing animal. The same sort of controlling system may be at work in zoothamnium, and it would be interesting if it could be identified.

THE SLIME MOLDS

The cellular slime molds, as the name implies, are somewhat less lovely than zoothamnium but even more interesting from the point of view of many biologists. The slime mold *Dictyostelium discoideum,* for example, begins life as a group of individual amoebas with no discernible connection with one another. Then, at a particular stage in their life cycle, these individuals quite suddenly swarm together and form a very metazoan-like creature. This creature then—and one can watch this happen—quite abruptly toadstools up into a plantlike structure, a fruiting body, bearing on the pinnacle of its long stalk one single shimmering droplet in which is contained hundreds or thousands of tiny spores. When these spores are dispersed, each one releases one small, quite ordinary-looking amoeba, and the entire cycle begins again.

The individual spore is a small cylinder that looks like a little

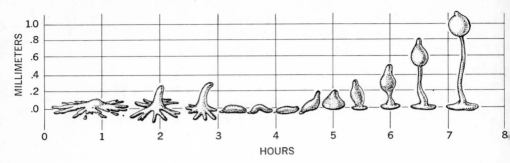

FIGURE 56 The life history of the cellular slime mold *Dictyostelium discoideum*. Hundreds of small amoebas swarm together, forming a sausage-shaped mass which eventually toadstools up into a fruiting body. On top of the stalk, which is made up of the bodies of amoebas, is a shining droplet containing spores. When these spores are dispersed, each one releases a new amoeba and the cycle begins anew. After Bonner.

gelatin capsule. This capsule is made of an outer layer of cellulose-like material and an inner layer of slime, two materials which are important in the rest of the cycle. In a warm, damp environment—the slime molds live on the damp earth or in decaying vegetation—the capsule splits open down its side and one amoeba emerges. This amoeba appears much like any other common soil amoeba, and in its way of life it is indistinguishable. It eats bacteria, grows, and multiplies rapidly until soon—within a day or two—most of the bacteria food supply is gone and there is a large population of small amoebas. Then quite suddenly a change comes over the colony. The cells lose their food vacuoles, decrease in size, and become more elongated. Then they begin to move toward one or a few centrally located amoebas. At first they move as independent cells but soon they begin flowing in streams, each amoeba sticking to the ones ahead, behind, and to the sides of it. Experimenters have shown that in almost any group of this species of amoebas a centrally located cell will begin to "call" the others to it. The compelling force is a chemical that has been named acrasin after Acrasia, the cruel witch in Spenser's *Fairie Queene* who attracted men and turned them into beasts. As the cells are bathed in this chemical they become sticky and in turn begin to exude acrasin, serving to attract outlying cells. Presumably, there is a higher concentration of acrasin in the center of the mass, however, for they continue to move toward the middle with increasing velocity, hustling each other along as on a busy one-way thoroughfare. If the center of one of these aggregating groups is scooped up and moved to the edge of the colony, the streams of traffic turn around—as if they had back and front ends—and flow toward the new center. Acrasin has been isolated recently, and preliminary reports suggest that it may be a chemical of a class known as the steroids, the same class of chemical to which our own potent sex and adrenal hormones belong.

The "beast" that the swarming amoebas turn into (although it is much smaller) very closely resembles a common garden slug, and it moves along slug-fashion with a small pointed "head," or apex, and a broader base, leaving behind it a track of slime. The "slug" is actually not a single animal but a mass of individual amoebas all stuck together but all vigorously moving their pseudopods, even the ones in the center of the sausage-shaped mass. The slime around them is a thin sheath, a sort of tunnel secreted by the amoebas through which they crawl, then leave behind them. The traction for the colony is not between colony and ground but between individual amoeboid cells and the inside of the sheath.

This slug, or sausage, behaves as if it were a single organism. In the dark it will move tip-first toward faint light or toward infinitesimal gradations of heat as if it were trying to reach a warm open spot ideal for scattering its progeny. This migration may last anywhere from five hours to two weeks. During this time, changes take place in the cells. First there are changes in position among the individual cells; these have been carefully studied and recorded because the cell position is of great importance in the events to come. We know that, at the time of aggregation, the cells are thoroughly mixed; if those in the attractive center are stained blue with a harmless dye, the beast when it forms is uniformly blue. In other words, the tip does not form as some experimenters had hypothesized, from the acrasin-producing center cells, which then continue to lead the sausage forward during its migrations. Rather, these cells become scattered throughout the slime mold. Once the amoebas are all in motion, some rearrange themselves, the faster ones going forward and the laggards dropping back. If a group of fast-moving cells are removed from the anterior portion of the slug and grafted onto the rear, they will slowly, as a group, move forward again to their original rank. Similarly, a slower-moving contingent, if transplanted, will soon drop back to its previous position.

One might think, having learned that cells shift their position during migration, that certain cells are predestined to play certain roles—base, stalk, spore—in the culmination phase of the cycle. But if any one of these little sausages is cut in three parts, the cells of each part will differentiate to form a small but perfect fruiting body. In other words, the factor that seems to control the role that the individual cells play is not any characteristic of the cell itself but its position in relation to others in the cell mass, as is also the case in the quite different organization of zoothamnium.

As the cell mass reaches the end of its migration, other changes begin to take place. The cells in the forward tip begin to get swollen and vacuolated; these will form the stalk of the fruiting body. The front end of the mass stops moving forward and the rearguard cells bunch themselves underneath it, forming a blob with a little tip on it. Then this tip, made up of swollen cells, turns and pushes downward toward the base through the rest of the cell mass until it reaches the bottom. At the same time a small group of cells, which were at the rear end of the moving sausage, are also becoming swollen and vacuolated. These are to form the circular base of the mature fruiting body, which is similar in shape and proportions to the foot of a tall-stemmed glass. The stalk cells manufacture a cellulose-like product, which is deposited outside the cells, forming a stiff cylinder. As this cylinder pushes downward on the base, the cells that are to form the spores are carried upward, circling the stalk like a minute ball impaled upon a pin. As they move upward into the air, they too undergo changes. Each becomes smaller, dry, and rounded, and secretes for itself a hard protective outer coat. Above the little sphere of spores-to-be, the remaining stalk cells climb or are pushed to the top of the cylinder, make their cellulose contribution to it, and drop inside, where they will dry out and die. When the last stalk cell has so sacrificed itself, the spore cells in their little shimmering drop of mucilage are at the very pinnacle of the stalk. This drop contains some 85 percent of the original cell mass, with only about 15 percent sacrificed to the goal of raising the spores where they can be dispersed through the new territory found by the migrating sausage. Once these spores fall onto the warm damp ground, each will hatch a new small amoeba, indistinguishable from other soil amoebas, but each containing coded within it the entire information for this whole remarkable sequence of events. The cellular slime mold never, in fact, becomes a single, totally viable organism; it cannot even feed in the course of its migration, but it is a spectacular instance of cell differentiation at this one stage of its life cycle.

VOLVOX

One of the most distinguished of the colonial protozoa is volvox, a hollow sphere made up, according to age and species, of 500 to 50,000 tiny cells. Unlike gonium, which is its close relative, volvox is a model of biological organization and economy. The individual cells in volvox, like those of gonium, closely resemble chlamydomonas, small and bright green, with two equal flagella and a tiny red eyespot. They are held together not only by a clear mucilage but also by fine protoplasmic strands that interconnect them, forming a hexagonal pattern over the surface of the sphere. Actually, one cannot ordinarily see these connections between the cells; thus, when volvox whirls through the water it appears like a spinning universe of individual stars fixed in an invisible firmament. Inside each universe are often smaller universes, the daughter colonies that spin with it.

Leeuwenhoek saw volvox, on August 30, 1698, in water taken from a ditch. ". . . on coming home, while I was busy looking at the multifarious very little animalcules a-swimming in this water, I saw floating and seeming to move of themselves, a great many green round particles of the bigness of sand-grains. . . . This was for me a pleasant sight because the little bodies aforesaid, how

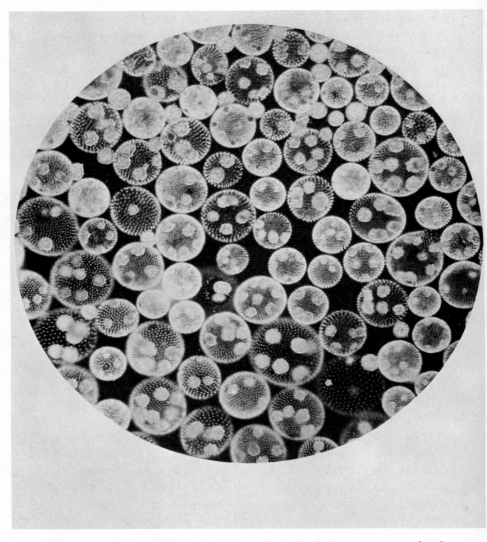

FIGURE 57 Volvox. Within the hollow spherical colonies, you can see daughter colonies and, if you look closely, you may be able to distinguish a third generation spinning within the second. (Up to 500 microns.) Photograph courtesy General Biological Supply House, Inc., Chicago.

oft soever I looked upon them, never lay still; and because too their progression was brought about by a rolling motion; and all the more because I imagined at first that they were animalcules. . . . Each of these little bodies had enclosed within it five, six, seven, nay some even twelve, very little round globules, in structure like to the body itself wherein they were contained. . . . Many people seeing these bodies a-moving in the water, might well swear they were little living animals; and more especially when you saw them going round first one way, and then t'other."

Volvox achieves this seemingly effortless coordination that Leeuwenhoek observed so well because each cell has a precise relationship to its neighbor that gives it a clearly defined though simple role. The main reason for this seems to be that it, like the slime molds, has a definite polarity—volvox has a top and bottom, permanent north and south poles. The flagella of each cell beat sideways in such a way as to spin the entire colony around this lateral axis. The movement of volvox is purposeful; it seeks the sunlight since it is photosynthetic, but avoids too strong a light. The entire sphere can stop and reverse itself, orienting to the light like euglena. Volvox kept in a jar or flask on the windowsill will tend to congregate in particular areas, moving its position as the light shifts, intensifies, and ebbs during the day. This movement to or away from the light is made possible because the cells on the too light or too dark side give an extra little kick with their flagella, a backward thrust that pushes the colony forward even as it spins.

If an isolated cell is cut loose from the colony by severing the threads that hold it to its neighbors, it is able to swim around like any other green flagellate, but with one important exception: it cannot reproduce. Reproduction takes place only within the colony and, for reasons which are not entirely understood, only within cells in the southern hemisphere. These cells, which appear identical in the young colony, become larger and greener—morphologically distinct—as the colony matures.

Reproduction is usually asexual. One of the southern cells enlarges and then begins to divide. The new daughter cells divide in turn but tend to stay small and closely packed, which is why the daughter colonies appear much darker than the parent. As they divide, held together by a sticky matrix and by protoplasmic threads, they remain attached to the original reproductive cell in the mother colony, forming a little pouch, or balloon, on the inner surface of the original sphere. At this stage, all of the new-formed cells are facing inward, their anterior ends turned toward the center of the sphere.

The next step in the development of the daughter volvox is so improbable that it was not observed by microscropists until only recently. At the point at which the daughter sphere is attached to the mother cell there is a small hole, an umbilicus, so to speak, which is known as the phialospore. Once the new colony is completed and is held to the mother cell only by remaining mucilaginous threads, it suddenly, and with remarkable coordination, turns itself completely inside out through the phialospore, like the finger of a glove. Then, the heads of the cells all pointing in the right outward direction, each cell sprouts flagella and the daughter is ready for independent existence. Daughter colonies remain inside the mother colony until the latter breaks apart to let them loose, and it is not unusual to see daughters with a third generation forming, or sometimes already completed, within them.

Volvox may also produce male cells, or gamonts; these too form only in the southern hemisphere. Depending on the species, some colonies produce only male cells, some only female cells, and some are hermaphroditic. Colonies of sperm cells are produced in much the same way as the asexual daughter colonies. One cell begins to enlarge and then to divide, developing into a little inward bulge of small cells. This, too, eventually turns itself inside out. The cells each develop two long flagella; then the completed bachelor colony pushes its way gently out of the parent colony. Not until then do the individual sperm cells separate. Meanwhile, female cells have been preparing themselves by growing larger. After penetration by the male cell, the fertilized female cell breaks away from the colony, develops a hard spiny outer coat, and drops to the bottom of the pond.

It is in this form that volvox survives the rigors of the winter, emerging each spring as it has for centuries to begin a new green universe. These spectacular colonies are not animals, as Leeuwenhoek thought they might be—not metazoa, in any way that the term is now defined. At best, they offer a suggestion of evolutionary clues as to how the metazoa came about—tenuous threads reaching farther back in time than our science can follow. Yet, the notion that such a creature, serene and majestic, spinning through the sunlight, is somehow a part of man's own heritage, is a pleasing—even marvelous—thought indeed.

BIBLIOGRAPHY

This bibliography is intended for readers for whom THE MARVEL-OUS ANIMALS is an introduction to the protozoa, although the books and articles on special subjects might be of interest to the more experienced reader as well. The book by the Jahns is indispensable for anyone who wants to actually look at any of the one-celled animals; an inexpensive spiral-bound taxonomic manual, it is lucidly organized to guide the reader to the identification of the types of protozoa he is most likely to see. Manwell's text is a thorough general treatment, agreeably written; it has little emphasis on taxonomy and there are several chapters devoted to the sporozoa. The most exhaustive book on the biology of the protozoa is the advanced text by Dogiel, which is particularly informative on questions of evolution and ecology. The book by Mackinnon and Hawes is devoted to extensive descriptions of a relatively few species, but these descriptions, particularly of the living animals, are extraordinarily lively and apt. The texts by Hall and Kudo are for the more advanced student. Hall's book, *Protozoology,* combines a general text with a taxonomic guide, while Kudo concentrates chiefly upon taxonomy, providing descriptions of some four thousand species. Hall's most recent book, *Protozoa,* which is a paperback, is particularly concerned with protozoan nutrition, a subject on which he is an expert.

Under the *General Zoology* heading, the book by Barnes is a good, well-written textbook with an introductory chapter on the protozoa. Buchsbaum and Milne is written for a more general audience and is noteworthy not only for the lively, lucid style that is characteristic of these authors but also for the spectacular photographs, both in color and in black and white. Bonner's *Cells and*

Societies contains three chapters on the protozoa, among its many other delights, and is recommended without qualification to all readers.

Most of the titles under *Special Subjects* are self-explanatory. They are all listed here first of all because every one of them is in itself worth reading; all of them contain material not covered at all or not nearly so extensively in the general texts; and lastly, because I am particularly indebted to these authors. The early brilliant book by Lwoff is rapidly becoming a classic in its field, perhaps of particular interest in view of his recently having been awarded the Nobel prize. As I glance over this list, I am happily struck, once again, with the extraordinary liveliness and sparkle of the writing of most protozoologists, which makes all excursions into this field even more felicitous and rewarding.

PROTOZOOLOGY

DOGIEL, V. A. (Revised by J. I. Poljanskij and E. M. Chejsin). *General Protozoology*. London: Oxford University Press, 1965.

HALL, R. P. *Protozoa*. New York: Holt, Rinehart & Winston, 1964.

—— *Protozoology*. New York: Prentice-Hall, 1953.

JAHN, THEODORE LOUIS and FRANCIS FLOED JAHN. *How To Know the Protozoa* (ed. by Harry E. Jaques). Dubuque, Iowa: W. C. Brown Co., 1949.

KUDO, RICHARD R. *Protozoology* (fifth edition). Springfield, Illinois: Charles C. Thomas, 1966.

MACKINNON, DORIS L. and R. S. J. HAWES. *Introduction to the Study of Protozoa*. London: Oxford University Press, 1961.

MANWELL, REGINALD D. *Introduction to Protozoology*. London: Edward Arnold Ltd., 1961.

GENERAL ZOOLOGY

BARNES, ROBERT D. *Invertebrate Zoology*. Philadelphia, W. B. Saunders, 1963.

BONNER, JOHN TYLER. *Cells and Societies*. Princeton, New Jersey: Princeton University Press, 1955.

BUCHSBAUM, RALPH and LORUS J. MILNE. *The Lower Animals: Living Invertebrates of the World.* New York: Doubleday & Co., Inc., 1962.

SPECIAL SUBJECTS

BONNER, JOHN TYLER. *The Cellular Slime Molds* (second revised edition). Princeton, New Jersey: Princeton University Press, 1966.
BOVEE, E. C. "Studies of feeding behavior of amebas. I. Ingestion of thecate rhizopods and flagellates by verrucosid amebas, particularly *Thecamoeba sphaeronucleolus.*" *Journal of Protozoology,* 7:55, 1961.
CLEVELAND, L. R. and A. V. GRIMSTONE. "The fine structure of the flagellate *Mixotricha paradoxa* and its associated microorganisms." Proc. Roy. Soc. B. 159: 668–86, 1964.
DOBELL, CLIFFORD. *Antony van Leeuwenhoek and His "Little Animals."* New York: Dover Publications, 1962.
FROUD, JOAN. "Observations on Hypotrichous Ciliates in the Genera Stichotricha and Chaetospira." *Quarterly Journal of Microscopical Science,* 90:141, 1949.
JENNINGS, HERBERT S. *Behavior of the Lower Organisms.* Bloomington, Indiana: Indiana University Press, 1962.
LWOFF, ANDRE. *Problems of Morphogenesis in Ciliates.* New York: John Wiley & Sons, Inc., 1950.
PITELKA, DOROTHY R. *Electron-microscopic Structure of Protozoa.* New York: Pergamon Press, 1963.
SUMMERS, F. M. "Some aspects of normal development in the colonial ciliate *Zoothamnium alternans*" *Biological Bulletin,* 74: 130, 1938.
TARTAR, VANCE. *Biology of Stentor.* New York: Pergamon Press, 1961.

INDEX

Acellular, proposed term for protozoa, 9

Acid, amino, 49; base balance of, 75; deoxyribonucleic, 51–52; and noctiluca, 98; ribonucleic, 52; stimuli for paramecium, 78–79; weak solutions of, 79

Acrasin, description of, 168–70

Actinophrys sol, 114

Actinosphaerium, description of, 69–70, 112–14

Adenosine triphosphate (ATP), 36

Adrenal hormones, 169

African sleeping sickness, 146–47, 149

Aging, of tokophrya, 62–63

Algae, 4, 23, 81, and amoeba, 109; blue-green, 36, 93; and chlorophyll, 5, 32; and chrysomonads, 96; as a food, 112–14; formation of, 91; green, 128; simple, 148; symbiotic, 83

Amoeba, 6, and acrasin, 169–70; and algae, 109; behavior of, 79–80, cell divisions, 55, discovery of, 1; emergence of, 168; explorations of, 15; feeding habits, 109–10; and forams, 117; formation of, 71; individual, 167; and molecules, 109; motion of, 14–15, 32, 109; and nutrition, 38; and parasites, 146, 148; primitiveness of, 107; pseudopods of, 112; quadrinucleated, 149; reproduction of, 56, 65; soil-dwelling, 37, 171; and spores, 171; varieties of, 15

Amoeba proteus, 36, 79–80, 109–10; behavior of, 83; cell division, 55; motion of, 14–15

Amoebic dysentery, 6, 149

Amoeboid, 104; and chrysomonads, 96; classes of, 27; and dinoflagellates, 98

Anaerobes, formation of, 92

Animalcules, 3, 7, 21, 130, 140, 171, 173

Animal life, 4, 93

Animals, cells of, 16; and chlorophyll, 32; and cirri halteria, 23, 25; control of, 23; domesticated, 15; family of, 106; heterogeneous, 7; higher, 10, 160; inbreeding among, 69; larval, 61; lower, 146; marine, 115, 147; modern, 160; multicelled, 7, 9, 32, 37, 51, 62, 67, 96, 159–60; one-celled 5–7, 9, 12, 27, 49, 51, 62, 70, 75, 84, 145, 154, 159–61; and sugar, 101; sun, 112, 114

Anopheles mosquitoes, and disease, 152–54

Apostomes, and parasites, 145

Aquariums, 91, 141

Arcella, description of, 55, 111–12

Aristotle, 121

Asexual reproduction, 65, 67, 70–71, 150

Atmosphere, 93

Atom, the, and energy, 31

Antogamy, and reproduction, 69–70

Autointoxication, and waste materials, 63

Axopods, of actionsphaerium, 112–14; feeding mechanism, 115

B_{12}, and euglena, 49; and vitamins, 30–31

Bacteria, and algae, 32; bioluminescent, 99; colored, 128; common, 104; in culture, 23, 48; and disease, 6; as a food, 36, 74, 79, 81, 109, 127, 136, 141, 148 168; formation of, 92–93; genetic recombinations of, 65; and gymnostomes, 124; habits of, 12; lineage of, 4, 29; and nutrition, 29, 30, 40; and paramecium, 7; and phagocytosis, 37; photosynthetic, 93

Behavior, patterns of, 75–76, 79–80, 104

Binary fission, 55, 67, 74

Biochemistry, 9, 30, 36, 49, 92

Biology, fundamental, 9, 11; rhythm of, 99, 101

HELENA CURTIS is a free-lance science writer who has written articles for encyclopedias, *American Scientist, Natural History, Nature and Science, Sea Frontiers* and *The Rockefeller University Review*. She has been a staff writer for the Sloan-Kettering Institute for Cancer Research and for Rockefeller University. In 1963, she received a Sloan-Rockefeller advanced science writing fellowship at Columbia University. Her first book, *The Viruses*, was published in 1965 by the Natural History Press. A graduate of Bryn Mawr College, Mrs. Curtis presently lives in Wood's Hole, Massachusetts.

33670

593
C978m

Curtis, Helena

The marvelous
animals

33670

593
C978m

Curtis, Helena

The marvelous
animals

DATE	BORROWER'S NAME	

593
C978m

33670

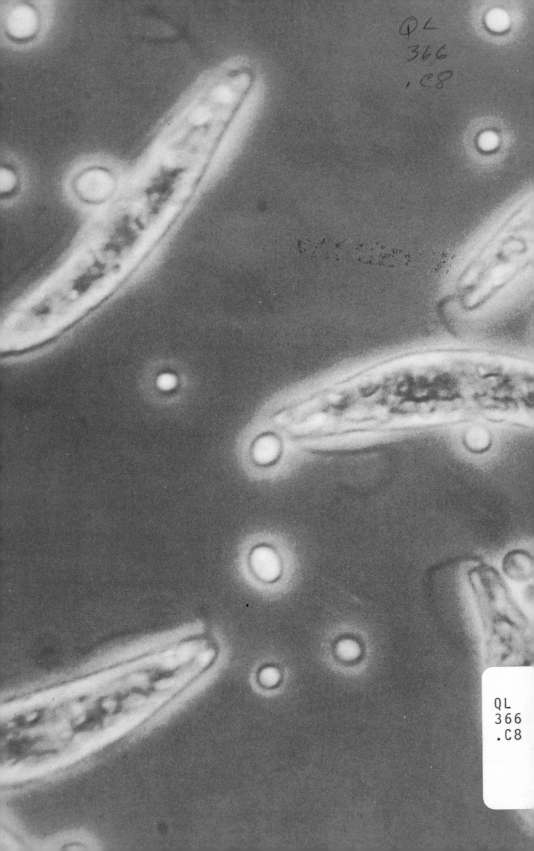